Praise for

The Emergent Agr

Tipping points in food and farming are obvious to anyone courageous enough to look. This highly readable treatise explains the healing future awaiting us. Let's learn, embrace, and move forward. This book describes the future perfectly.

—JOEL SALATIN, Polyface Farm

Here's the inside story on the most hopeful development in American culture in recent years. As local food systems grow and intertwine, they form a subversive challenge to the too-big-to-fail agriculture we've somehow come to accept as normal.

—BILL MCKIBBEN, author *Deep Economy*

In *The Emergent Agriculture,* Gary Kleppel leverages his extensive experience as an ecologist, teacher and farmer for an honest, sound and accessible examination of the often hidden costs of our current industrial agro-food system and the rise of one based on ethics, ecology and community that's replacing it.

—SEAN CLARK, farm director and agricultural ecologist,
Berea College, Kentucky

Dr. Kleppel makes an eloquent and well-researched case for supporting agricultural production that is community-based, appropriately scaled to local resources, nutritionally rich and ethical. The content evokes Wendell Berry and Michael Pollan, but with many fresh insights about the enriched food systems that are sprouting from the ground up across the globe.

—MARIANNE SARRANTONIO, Associate Professor of
Sustainable Agriculture, University of Maine

THE EMERGENT AGRICULTURE

FARMING, SUSTAINABILITY AND THE RETURN OF THE LOCAL ECONOMY

GARY KLEPPEL

FOREWORD BY JOHN IKERD

Cover design by Diane McIntosh.
Cover images ©iStock (Horse and farmer: small frog, frolicking lambs, Mathew Lees Dixon; organic farmer: alle12)

Printed in Canada. First printing June 2014.

New Society Publishers acknowledges the financial support of the Government of Canada through the Canada Book Fund (CBF) for our publishing activities.

Paperback ISBN: 978-0-86571-773-2

eISBN: 978-1-55092-577-7

Inquiries regarding requests to reprint all or part of *The Emergent Agriculture* should be addressed to New Society Publishers at the address below.

To order directly from the publishers, please call toll-free (North America) 1-800-567-6772, or order online at www.newsociety.com

Any other inquiries can be directed by mail to:

New Society Publishers
P.O. Box 189, Gabriola Island, BC V0R 1X0, Canada
(250) 247-9737

New Society Publishers' mission is to publish books that contribute in fundamental ways to building an ecologically sustainable and just society, and to do so with the least possible impact on the environment, in a manner that models this vision. We are committed to doing this not just through education, but through action. The interior pages of our bound books are printed on Forest Stewardship Council®-registered acid-free paper that is **100% post-consumer recycled** (100% old growth forest-free), processed chlorine-free, and printed with vegetable-based, low-VOC inks, with covers produced using FSC®-registered stock. New Society also works to reduce its carbon footprint, and purchases carbon offsets based on an annual audit to ensure a carbon neutral footprint. For further information, or to browse our full list of books and purchase securely, visit our website at: www.newsociety.com

Library and Archives Canada Cataloguing in Publication

Kleppel, G. S. (Gary S.), author

The emergent agriculture : farming, sustainability and the return of the local economy / Gary S. Kleppel ; foreword by John Ikerd.

Includes bibliographical references and index.
Issued in print and electronic formats.
ISBN 978-0-86571-773-2 (pbk.).--ISBN 978-1-55092-577-7 (ebook)

1. Sustainable agriculture. 2. Sustainable agriculture--Economic aspects. 3. Farms, Small. 4. Farms, Small--Economic aspects. I. Title.

S494.5.S86K64 2014 333.76'16 C2014-902321-9
C2014-902322-7

For Pam and Jarret

Contents

Acknowledgments

THIS BOOK WAS A WORK OF PASSION. It emerged as a result of the contact I have had over the past decade with some of the most amazing people I have ever met — the people who produce our food. I will be forever grateful to the farmers and food service people who allowed me to tell their stories and relate vignettes and anecdotes about their lives. *The Emergent Agriculture* took shape as I told the stories of people like Jim and Adele Hayes, Mark and Kristin Kimball, the Ball family, the Abbruzzese family, Shannon Hayes and Bob Hooper, Thomas Christenfeld, Slim and Cathy Newcomb, Kathleen Harris, Jean-Paul Courtens and Jody Bolluyt, Jody and Luisa Somers, Stephen Pearse, and Dan and Vicki Purdy. Many others, not mentioned by name within these pages, have been just as inspirational. These include, Barbara and Bernie Armata, Regina, Bill and Christian Embler, Alexander "Sandy" Gordon, Ken Kleinpeter, Richie Gaige, Jr. and Amanda Terhune, Seth McEachron, Ben and Lindsey Shute, Clemens Mackay and Jenny Rosinki, Jon and April Audietis, and Severine von Tscharner Fleming. These people epitomize, in the way they live, work, and relate to others, the iconic image of the American farmer. They work so hard and set such a high standard of excellence in the pursuit of their craft and in their ethics that it is impossible not to notice them, respect them, and want to follow the example they have set. These folks have been good and patient teachers, superb role models, and clear examples of the best that America has to offer. And yet, they are typical of farmers everywhere. They are the reason I wrote this book.

Although I have been publishing in the technical and discipline-based literature for nearly 40 years, writing socially-focused non-fiction proved surprisingly tough and sometimes humbling. As I read and re-read drafts of the manuscript, I realized how difficult it is to communicate in "my own voice." Syntax, simple grammar and the creation of sentences on the page that "sounded" as though I was speaking directly to the reader, proved challenging to say the least. I owe an enormous debt of gratitude to Paul Tick, and to Rebecca and Roger Meyers — all of whom I met at the Delmar Farmers' Market — for their editorial assistance. I am also grateful to Ian le Cheminant, my copy editor, who with nearly the patience of a saint managed to get the final draft of the manuscript ready for press, almost on time. I am indebted as well to my friend and teacher, Dr. Richard Pieper, emeritus of the University of Southern California. Thirty years ago, Rick, an oceanographer, was my post-doctoral advisor. He read my work then — as an expert in my discipline — and he read my work now, as a typical consumer of literature, but without the background in agriculture. Now, as then, he let me know what made sense and what didn't, and where I needed to modify the text to reach an audience that was not assumed to have expertise in farming or to be familiar with the message of the book.

Finally, there is the dog. In a recent article, The New York Times referred to the border collie as the farmer's most useful tool. I can attest to that. I would probably not still be farming if not for my border collie, Tory. There are probably people reading this who see little point acknowledging the contribution of a dog to one's book. I would have to agree. It is hokey. And Tory, as smart as she is, won't be reading this anytime soon. So what's the point?

The point is simply that this book would not have been written had Tory not been at my side as I learned to manage sheep, as I took the concepts and techniques that I read about in books and learned about during seminars, and turned them into my personal pasture management approach. When Tory joined our farm she was four years old, fully trained (by Barbara Armata) and had more experience with sheep than Pam and I put together. As I learned the first time I tried to move sheep without a dog, intensive rotational grazing is impossible, or nearly so, without the negotiating power provided by a pair of large canines. Tory has made it possibly for me to raise sheep in a manner I consider sustainable. She is as vital to our operation as Pam or I. And Tory adds something else — something lost in the industrial age and returned to me each time we take to the field — the ability to communicate across species. The smooth (or sometimes not so smooth) rotation of a few dozen sheep from one paddock to the next with the help of a dog is a beautiful thing. It is one of those experiences one never tires of, that makes the work enjoyable and sometimes thrilling, even when the day is hot and humid and a few acres of fencing need to be moved. Watching my dog work deepens my understanding of the meaning of the words loyalty and enthusiasm, of intelligence and athleticism. That is my dog. She makes it possible for me to do my work. Even if she can't read these pages, you can. And you should know who she is and the vital role she plays in the life of a farmer.

Foreword

By John Ikerd

A NEW KIND OF AGRICULTURE IS EMERGING IN AMERICA. This may come as a surprise to those who are caught up in the American food culture of "quick, convenient, and cheap." Talk of a quiet revolution has been common in sustainable agriculture circles for at least the last 25 years. Hopefully, the "emergent agriculture" which Gary Kleppel portrays so eloquently in this book has brought us to the cusp of creating a new vision of the future of food, one so compelling that few will be able to resist joining the new food and farming revolution.

This is a book of stories. As Kleppel writes, he has found stories to be more effective than lectures in reaching the minds and hearts of his students. Following his lead, perhaps my personal story will shed some additional light on the prospects for revolutionary change in our food system and a return of local economies. One advantage of being old is that I can remember when things were quite different from today. I have lived through a fundamental change in our food system, and I know the future of food will be very different from either the present or past.

I grew up on a small farm in the southwest Missouri Ozarks during the '40s and '50s. During my early years, we had neither electricity nor indoor plumbing. We kept a few hogs and chickens and a large garden that met much of our food needs. We depended on a small herd of dairy cows, milked by hand, to provide us with milk and the money we needed to buy the rest. There were no supermarkets or franchised restaurants in my world as a kid. My mother would take her grocery list to the nearby general store and hand it across the counter to the proprietor, who would select her items from the shelves, barrels, or meat case, and place them in a brown paper bag. The cost would be added up and either paid or "put on the bill." Nearly all of the food in that store came from within 50 miles of our home — notable exceptions being coffee, spices, and an occasional stalk of bananas. The only restaurants in the nearby towns were locally-owned, family-operated cafes and truck stops.

I remember when I first heard of the opening of a Piggly Wiggly (a national chain supermarket) in a nearby town. For the first time, people were going to be allowed to select their own items from the shelves. This was the beginning of a revolution in food retailing that would still take some time, but would eventually lead to today's food mega-markets. I saw my first McDonald's while in college in the late '50s. This was the beginning of the "fast food" revolution. During the '60s and '70s, corporate chains of supermarkets replaced the "mom and pop" grocery stores, franchised restaurants replaced the local cafes. The industrial food revolution emerged into full view.

I can also remember a time when the teachers in my two-room grade school let us kids go outside and watch a steam engine moving slowly down the road on its way from one farm to the next. The engine pulled a large threshing machine. Once it was set up at the next farm, the steam engine would also power the threshing operation, mechanically separating grain from straw as it processed the neat shocks of sheaves that had been stacked by hand and left to dry in the fields. This was the beginning of a revolution in farming. But it would take some time to transform American agriculture. After World War II, steam engines and horses were quickly replaced by tractors, chemical fertilizers replaced livestock manure, pesticides replaced crop rotations for controlling insects and weeds, and the industrial agricultural revolution emerged into full view. In less than 40 years, agriculture was transformed into agribusiness.

Perhaps most relevant to today's emergent agriculture, I remember the threshing crews that brought dozens of local farmers together to share in the work of cutting, shocking, and threshing grain. Other crews of local farmers of various numbers were formed to fill silos, bale hay, and kill hogs. The farmers' wives and daughters shared in the task of harvest by preparing dinner, the noonday meal, for those crews, at times ranging up to forty hungry men and boys. Those with the largest farms typically provided the most expensive equipment while others provided horses, later tractors, wagons, or just family labor, to make up "their share" for the task at hand. The tasks on a given farm were not necessarily equal to that farmer's contributions, but no one seemed to question the size of either the task or the contribution. Money only rarely changed hands. These were not just families who worked together; they were people who lived together, forming close-knit farming communities.

It took a community to operate a farm in those days. Farmers couldn't afford to let conflicts arising from their work alienate or isolate them from their communities. Their relationships were relationships of necessity, not necessarily relationships of choice. They didn't complain when a neighbor's grain was threshed a day before their own, even when a big rain or hailstorm was predicted for the following day. They couldn't afford the luxury of disagreements over politics that might lead to a breakup of their threshing crew. It is not surprising then that many farmers in those days longed for the independence and freedom that owning their own grain combine, hay bailer, or silage chopper would eventually bring.

The industrial farming revolution would not only allow farmers to produce more at lower costs, and hopefully make more money, it would free farmers to operate their farm without depending on so many others in their communities. Farmers of the '40s and '50s had no way of knowing how much the independence and freedom brought by industrial agriculture ultimately would cost them — ecologically, socially, and economically. They didn't foresee the negative consequence for the health of the land or for local communities and local economies.

Many people equate industrialization with the migration of people from farms and rural communities to find manufacturing jobs in urban areas. However, the shift from agrarianism to urbanization is only one symptom of the industrial model or paradigm of economic development, which is characterized by specialization, standardization, and consolidation of control. Specialization increases economic efficiency: People who specialize in fewer things tend to do them faster and better. Standardization is then necessary to facilitate coordination, routinization, and mechanization of specialized production processes. Standardization simplifies production and management processes, allowing consolidation of control into large-scaled, eventually corporately-controlled, business enterprises. This is the basic process by which industrial organizations realize "economies of scale" — in agriculture as well as manufacturing.

The process of industrialization first came to manufacturing, but it is no different in concept from the industrialization of agriculture and the rest of the food system. In fact, the industrialization of agriculture facilitated the industrialization of much of the rest of the economy. As farms became larger, the number of farms and farmers necessarily became fewer, providing workers for the factories and offices of a growing industrial economy. The era of manufacturing in America may be over, but the industrial paradigm of organization still permeates and dominates virtually every aspect of our economy and society, including American agriculture.

Industrial agriculture brought tremendous increases in economic efficiency but it also brought many unintended environmental, social, and economic consequences. It is inherently dependent on non-renewable fossil energy, chemically-dependent monoculture cropping systems, and large-scale confinement animal feeding operations. We see the environmental consequences of these dependencies in eroded and degraded soils, polluted streams and groundwater, depleted streams and aquifers, and the growing threat of global climate change.

We see the social and economic consequences of industrial agriculture in the demise of independent family farms and the social and economic decay of rural communities as the farms have grown larger in size, fewer in number, and increasingly corporate-controlled. In addition, the most basic human rights of self-determination and self-defense are systematically denied to rural residents who are forced to live with the risks to public health increasingly linked to large-scale "factory farms."

These consequences are systemic. Everything of economic value ultimately comes from the earth, and beyond self-sufficiency comes by way of economic relationships within society. Industrialization is a very efficient means of extracting economic value from natural and human resources but it provides no incentives to protect or renew the resources for the benefit of society in general or the future of humanity. Nowhere is the critical nature of these relationships more clear and compelling than the sustainability of food production, which obviously depends on the productivity of the land and the competency, commitment, and integrity of those who take care of the land. In the pursuit of economic efficiency, industrial agriculture is destroying the economic foundation of rural America.

Perhaps most important and least understood, the industrial food system has failed in its most fundamental purpose of providing domestic food security. A larger percentage of people in the U.S. are "food insecure" today than during the 1960s, with more than 20 percent of U.S. children living on the verge of hunger in food insecure homes. In addition, the only foods affordable to many lower-income families are high in calories and lacking in essential nutrients, leading to an epidemic of obesity and related health problems, such as diabetes, heart disease, hypertension, and various forms of cancer. The irresponsible use of agricultural chemicals, growth hormones, antibiotics, and a multitude of additives in industrial foods add to a growing list of diet-related illnesses. We can no longer afford the high costs of cheap food.

Advocates and opponents quibble about the specific statistics and the extent to which problems with our current food system are linked to industrial agriculture, but there can be little doubt about its lack of sustainability. It is not meeting the needs of many people today, even in the United States, and it most certainly is not leaving equal or better opportunities for future generations. It is not ecologically, socially, or economically sustainable.

It was very difficult for me to come to this conclusion. During my undergraduate and graduate education in agricultural economics at the University of Missouri, I was taught and I believed that specialization, standardization, and consolidation into larger farms would be good for farmers and rural communities and would provide food security for American consumers. There would be profit for farmers who become more efficient, which would help build strong local economies. Ultimately, the benefits of greater economic efficiency would be passed on to consumers through greater quantities and lower prices for food. I was going to help make good food affordable and accessible for everyone.

Between work on my BS and MS degrees, I participated directly in the industrial food system by working three years for Wilson Packing Company, Inc. (now Wilson Foods, Inc.), the fourth largest meat packer in the country at the time. After returning to graduate school and completing my PhD degree, I spent the first half of my 30-year academic career working with farmers to increase the economic efficiency of their farming operations. I advised farmers that the only farms with a future were those managed for the economic bottom line. Modern farming was a business, not a way of life. Family matters couldn't be allowed to interfere with the farm business. Farmers needed to be prepared to either "get big or get out." Obviously, I was wrong, but it wasn't easy to admit that at the time.

My personal revolution came during the farm financial crisis of the 1980s. Network news programs at the time regularly featured farm bankruptcies, foreclosures, and occasional farmer suicides. As an agricultural economist at the University of Georgia at the time, I was responsible for trying to help Georgia farmers save their farms, or at least their lives. After working with a number of farmers during those difficult times, I eventually was forced to conclude that the farmers with the greatest financial difficulties were those who had followed the recommendation of the so-called agricultural experts, including myself. It should have been obvious that in order for farms to become bigger, the number of farmers had to become fewer, meaning that some farmers had to fail so others could farm their land. It wasn't the farmers' fault; failure was an essential aspect of the system.

The industrial paradigm of agriculture that I was taught and had been teaching certainly was not working and wasn't going to work for farmers, even the so-called "good farmers." I could also see it wasn't working for the rural communities that depended on farm families, not just economically but socially and culturally. I eventually understood that it wasn't working for those of future generations — and I came to understand even later that it also wasn't working for consumers. Industrial agriculture was then and still is destroying the productivity of the land and degrading the occupation of farming. It was not sustainable. I have spent my time and energy since the '80s trying to understand, educate, and advocate for agricultural sustainability.

During this time, the organic and local food movements have emerged and grown in direct response to a large and growing list of public concerns about the ecological, social, and economic integrity of the industrial food system. The modern organic movement arose from these concerns in the '60s but didn't gain widespread support until the sustainable agriculture movement emerged in the '80s. Organic food sales in the U.S. grew rapidly during the '90s and early 21st century, averaging 20 percent-plus per year and doubling every three to four years before stabilizing at around 10 percent per year during the Great Recession. National organic standards opened up organic production and distribution to large specialized farming operations and mainstream supermarkets. By 2007, the supermarkets and large natural food chains, such as Whole Foods and Trader Joe's, had gained dominance over organic food distribution and retailing. Large specialized "industrial organic farms" emerged to meet the demands of mainstream supermarkets.

As organic production moved to larger farms and into mainstream markets, organic consumers increasingly looked to farmers in their own communities to ensure the ecological and social integrity of their food. Kleppel documents the emergence and growth of the local food movement very effectively in his book, so I won't dwell on the movement here. I simply wanted to endorse his conclusion that: "People want the products that sustainable agriculture has to offer. People want good tasting, clean, nutritious, safe food. People want to support their local economies. People want to know that what they eat was produced humanely. People want farms in their neighborhoods, regions, states, and nation. People want to know that the iconic perception of American agriculture actually describes the system that produces their food." People increasingly are looking to local farmers to produce their food because of their growing realization of the failure of the industrial paradigm of agriculture.

I believe the future potential for the emergent sustainable food system can be seen most clearly in the growing number of local food cooperatives or collaborations among farmers and their customers. Some of these collaborations began as multi-farm community supported agricultural organizations or CSAs, where two or more farmers share the responsibility of providing a variety of fresh produce for their CSA members on a weekly basis throughout the growing season. By including additional farmers in their collaborative, they can offer a wider variety of local products — meat, milk, eggs, honey, and other local products. They can offer purchase-and-delivery options ranging from CSA shares and standing orders to on-line orders of individual items.

The USDA Agricultural Marketing Service currently lists over 230 multi-farm "food hubs" that offer a wide variety of products and services. These new community-based food systems range in size from a dozen or so to hundreds of farmer/consumer members. The sustainability of cooperative ventures or alliances does not depend on their organizational structure but instead on their commitment to shared social and ethical values, which can be sustained only through positive personal relationships.

Kleppel provides an overview of the various sections of his book in his introduction. So, I will focus my remaining comments on what I believe is its seminal contribution to the sustainability of farming and the local food movement: *The importance of personal relationships*. He tells insightful and compelling stories of emergent farmers who are building personal relationships with their customers, not just to create profitable markets but also to form purely personal friendships. These farmers and their customers are finding a renewed sense of connectedness or community through farmers' markets, CSAs, and other direct marketing venues where they meet to transact business but also to cultivate friendships. These farmers give priority to people in their local communities in marketing their products and their customers give priority to their local farmers. Both farmers and their customers value the friendships they form and maintain through their common interest in good food.

The integrity of such relationships is critical to the sustainability of the food system. The sustainability of food freshness, flavor, safety, and nutrition, and of economic and social benefits for communities and societies all depend on sustaining the integrity of relationships among farmers, customers, and society — and their relationships with the earth. Relationships of integrity — creating and maintaining them — will be the greatest and most important challenge in transforming the current local food movement into a new sustainable food system for the future. This will not be an easy task in a society that has been splintered by the quest for personal independence and economic competition in the pursuit of ever-greater economic efficiency.

The personal relationships being formed around a common interest in good food, in both rural and urban communities, are reminiscent of the relationships among the farmers and their families found in the threshing crews, silo filling crews, and haying crews of my childhood. However, there is a fundamental difference. The relationships of the agrarian past were relationships of necessity. The relationships of our sustainable future must be relationships of choice. The emergent farmers of today are not farmers by necessity, they are farmers by choice. Most of them could be making more money in any number of alternative occupations. They farm "because they want to," because farming gives purpose and meaning to their lives. The local food customers of today do not buy from local farmers because they can't afford to buy food at Walmart or even Whole Foods; they buy from local farmers "because they want to," because they care about the integrity of their food and their farmer. New communities of choice — communities of common interests, values, and place — are being created by the sustainable food revolution.

The people in the emergent food system are forming relationships because they realize we are social beings and need relationships for reasons that have nothing to do with any economic value they may or may not receive in return. We humans need to care and be cared for, to love and be loved. We are also ethical and moral beings who need to feel a sense of rightness and goodness in what we do — to give our lives purpose and meaning. As social relationships become less personal, people begin to understand that we should express the same values in relating to people we don't know, including those of future generations, as we express in our relationships with our friends and neighbors. We will come to understand that the value of relationships arises not just from personal connections within our families and communities, but also from being members of societies and of humanity. This is the process by which personal social values evolve into a sense of community, patriotism, and eventually into impersonal ethical values.

As Kleppel writes: "When the importance of connectivity and relationships is appreciated, a stabilizing farm community and, ultimately, a network of communities emerges. With few exceptions people are hungry for these relationships. They are sustained by these relationships and they prosper from them. The emergent agriculture celebrates ethical relationships between farmers and consumers, farmers and livestock, farmers and the land." The social relationships depicted in the stories in his book evolve into ethical commitments of responsibility, equity, and justice for those of future generations.

Since ethical values are impersonal, they can also be extended to relationships with non-persons — to other species, air, water, forests, lakes, streams, rocks, to the earth. Stewardship of nature is an ethic that evolves out of direct human relationships with nature, as between farmers and the land they farm. Caring for the land is inseparable from caring for those of future generations. Concerns for the whole of society and the future of humanity are neither economic nor purely social in nature but are fundamentally ethical. Kleppel calls ethics the "third bottom-line" of sustainable agriculture. "It is the bottom line that emerges when one seeks to answer the simple question, 'What's the right thing to do?'"

This kind of thinking is certainly not consistent with the economic way of thinking that was taught in college and taught to farmers during my early academic career. However, it is the kind of revolutionary thinking that is absolutely necessary to support the sustainable food revolution. As Kleppel writes, "To be sure, the revolution is further along in certain places than others, but it is gaining momentum everywhere. Farmers' markets and other forms of direct marketing represent, nationally, the fastest growing sector in agriculture. Increasing numbers of consumers are seeking humanely and naturally produced food. Small scale organic production, the local food movement, and farm-to-table culinary venues are proliferating."

"Is it a fad? Time will tell, but I would bet it's not," Kleppel writes. The total sales of all "non-industrial" foods probably still amount to something less than 10 percent of total retail food sales in the US. However, the natural/organic food movement is less than 50 years old and was virtually unknown until about 25 years ago. The sustainable agriculture movement is much further along than the industrial agriculture movement was when I was a boy back in the '40s and '50s. Within 30 years, the food system was fundamentally changed. The "non-industrial" food movement has grown tremendously over the past two decades and continues to grow, in spite of a lingering economic recession. Within 30 years, the industrial food system will be a thing of the past — hopefully, replaced by a new sustainable food system. But that will be up to us.

Hopefully, the right spark at the right time will ignite a virtual explosion in public demand for good food — meaning food that is safe, healthful, nutritious, affordable, but not cheap, that is produced by thoughtful caring farmers with ecological and social integrity. It only takes a spark to reawaken our inherent need for connectedness with each other and with the earth. That spark can be kindled into a flame as we reconnect through our common interest in good food. It remains to be seen if that flame will explode into revolutionary change in our food system, but if and when the revolution comes, the emergent farmers will be ready. Regardless, Gary Kleppel's emergent agriculture is moving us ever closer to a revolution in food and farming and a return of local communities and economies driven by a quest for agricultural sustainability.

Introduction — On the Cusp of a Revolution

If you had lived in Boston, Massachusetts on March 6, 1770, it is likely that you would have been shocked by the incidents of the previous day. To increase control over its American colonies and their commerce, and to help pay the cost of years of war in Europe, the British Parliament had sought to restrict trade both within and outside the colonies and had levied taxes on goods and services important to the colonial economy.[1] Colonials found these acts, passed without their representation in Parliament, an abrogation of their rights as British citizens and many took to the streets in protest. On March 5, British soldiers, quartered in Boston, had opened fire on a mob of taunting citizens, killing five of them and wounding six. The incident would become known as the Boston Massacre. Despite your anger, however, you probably would never seriously have considered the possibility that the colonies were on the cusp of a revolution. Nor would you have thought that just six years later 56 wealthy, well educated men in Philadelphia, representing the two and a half million British subjects in America, would declare the independence of the thirteen colonies from Great Britain and, by proxy, war on the strongest military power on earth. You would certainly never have imagined that in just eleven years the colonies would have won that war, that Massachusetts would be part of the sovereign United States of America, and that you would be a citizen of that new nation.

Fast forward to the 21st century: If you visit a farmers' market every Saturday, or belong to a community supported agriculture (CSA) program, or make an effort to buy locally produced food, it might never occur to you that today we also are on the cusp of a revolution — a revolution in the way food is produced, marketed, and distributed in this country. But we are. I doubt that you feel much like a revolutionary as you stroll among the colorful tents overflowing with freshly picked carrots, leeks, and onions; tempting cheeses; and coolers full of pasture-raised meats and eggs. But as sure as the citizens' demonstration that led to the Boston Massacre was a step toward the American Revolution, so is your patronage of the local food system a step toward a revolution in American agriculture.

The comparison of the American Revolution with today's shifting paradigm in food production may at first seem rather off balance. Comparing the founding of our nation to the way we grow food may seem like an inappropriate analogy, at best, but it makes a point. The American Revolution had enormous political and socio-economic impacts. So does the modern revolution in agriculture. The American Revolution changed the American peoples' lifestyles and quality of life. Today's shifting agricultural paradigm is doing much the same thing. It is changing the way we produce food, how and where we get it, and what it does to our health, our pocket books, and our social lives. This new agriculture is changing the amount and kinds of energy used to produce food, and even the way our landscapes are structured. The changes taking place in our food system today, like those that gave rise to our nation more than two centuries ago, are nothing less than transformative. But while the American Revolution transformed our political and socio-economic systems, the current revolution in agriculture is transforming our life support system. We are, without doubt, in the beginning stages of an emergence.

An emergence is a sudden, universal transformation in the status quo that arises from numerous local transformations with similar outcomes.[2] Suppose, for example, that someone watching a baseball game in a crowded stadium feels a raindrop. She pops open an umbrella. Several more drops; more umbrellas. Pretty soon the stadium is a sea of umbrellas. That is how an emergence works. Let's say that the rain during the ball game is a metaphor for the industrial system of food production, the system that produces most of the food that Americans eat today, and that opening an umbrella represents a response leading to an alternative food production system.

This alternative food system, the emergent agriculture, is changing the way we produce and distribute food in this country. What began as a passion for hippies, health food junkies, and — interestingly enough — chefs from five star restaurants, has evolved into a mainstream American movement. In less than 20 years the number of farmers' markets has risen from fewer than 1,800 to more than 8,000 — an increase of almost 350 percent.[3] Similarly, in 1986, there were only two community supported agriculture programs in the United States. Today, there are over 6,000 — a 3,000-fold increase in less than 30 years.[4] That's an emergence!

In this book I am going to describe how this new food production and distribution system is changing not only the way we eat, but the quality of our lives. I'm going to do this from the combined personal perspectives of a farmer who is also an ecologist. I live on a small farm 15 miles west of Albany, New York where my wife and I raise sheep and chickens, and bake artisanal breads. In my other "day job" I am a professor of biology (ecology) at the State University of New York at Albany. From these two vantage points I see farming not simply from the perspective of food production, but from its more general relationship to the natural world and to the place that humans and agriculture occupy in that world. Agriculture is simultaneously affecting and being affected by a larger universe, elements of which include a rapidly changing climate and an unpredictable economic environment. To prevail, as agriculture must if our society is to survive these uncertain times, we must embrace a new set of principles — a new ethic — grounded in sustainability. This transformation is already happening, and its momentum is building. *The Emergent Agriculture* will take you on a journey from where agriculture is now to where it is going, and you will gain a clearer understanding why it is so important to get there.

The journey is comprised of 14 essays, divided among four sections. The essays in each section explore different elements of the food system. In the end, I believe you will agree that the current, dangerous, inhumane, and dehumanizing American industrial food production system is being replaced by something much better.

The first section of the book is devoted to a brief history of the industrial food system and what it has become. I contrast this with the emerging agricultural system and what that system has to offer, not simply to the market, but to the quality of our lives.

In the second section, I consider sustainability as a concept and as a process. I examine the relationship of farms and farmers to diversity, both biological and economic, and I describe some of the ways that diverse systems respond to perturbation, both natural and social. The final essays in this section examine the implications of two major environmental modifiers, climate change and fossil-based energy, on our ability to produce food, and even on our ability to survive as a species.

The third section focuses on economy, and the new economic and market models that are emerging in parallel to the changing paradigm in agriculture. I emphasize the critical role of the consumer in the emergent system and the extent to which the relationship between producer and consumer creates bonds that are socially and economically stabilizing, even in the face of environmental and political instability.

The fourth and final section of the book contains only one essay — *The Emergent Agriculture.* This piece takes you full circle. It summarizes the major ideas of the book while at the same time helping you to visualize the future of food production in America.

I found the voice for this book in my classrooms. I have been teaching college for nearly 30 years. If I deliver my lectures as bland narratives my students will take notes, memorize what they think will be on the exam, regurgitate that information when asked to, and then promptly forget everything. If I abandon formality and simply have a conversation with my students, relating personal stories and anecdotes to the topic of the lecture, students become engaged, and they remember more of what I've tried to communicate long after the exam. So I've learned to have conversations and to tell stories. *The Emergent Agriculture* is a conversation with the reader. It is full of stories and anecdotes. Some recount my experiences farming and describe how agriculture has shaped my understanding of sustainability. Others are about my friends and colleagues, their struggles, and their triumphs as they practice their crafts under increasingly unpredictable environmental and economic conditions. And others describe how the devotion of my university's students, faculty, and administrators to the concept of local food sourcing has changed the culture of the university's food service, connected farmers with our institution, and helped to answer the question "Can we feed everyone in America locally?"

The Emergent Agriculture documents the current revolution in food production and predicts where that revolution is taking us. It is an appreciation for the strength and resilience of our nation's farmers and an explanation for the iconic place that the family farm occupies in our vision of what America is and ought to be.

Part I

Farming: An Emerging Paradigm

PAM KLEPPEL

Really, the only two things that any of us have are land and labor.[1]

— Carter, Swancy, Riverview Farm, GA

Farming ten thousand acres… they really don't want to be doing that. At that point, you're a manager of some kind… If a man can't make a living off a couple of hundred acres of good land, the system's broke.[2]

— Wes Swancy (Carter Swancy's son), Riverview Farm, GA

1

A New Approach to Agriculture

It's three-thirty on a Monday morning in mid-May. I'm sitting in a ladder-back chair at the low wooden counter in my kitchen where I keep my farm records, sipping my second cup of coffee and, as on most mornings, working on one of my projects. I use this quiet time each day to analyze data, to prepare lectures for my classes, and to write. Today, I am working on this book. In about an hour-and-a-half, the dogs will come charging down the stairs from our bedroom, ready to begin their day. Pam and I will have breakfast and then I'll head out to the south pasture behind the barn and start pulling up the light-weight plastic fencing that I use to create temporary enclosures, or paddocks, that confine 20 or so ewes and lambs to about a quarter of an acre of land. The sheep have been grazing this quarter-acre for three days, and today I will move the fencing to an ungrazed section of pasture, creating a new paddock full of fresh, clean grass for the ewes and their lambs to feast on for the next three days.

At about the same time this morning, Jim Hayes, after joining his wife Adele for a cup of coffee, will leave his house at Sap Bush Hollow Farm, in New York's Schoharie Valley. He and his border collie will walk in the dim light to a small shed attached to the barn. As Jim opens the door, the amber light from a pair of heat lamps will reveal an explosion of chicks scattering wildly through the wood shavings that cover the shed floor, in a frenzied search for breakfast. In only eight weeks, these tiny birds, each weighing just a few ounces, will check in at five or six pounds and be ready for market. On this morning, Jim will feed the chicks, and then the layers and the hogs, and finally, crossing the road to a paddock in a pasture that he rents from a neighbor, Jim and his dog will move his sheep down the road to a new paddock, in another rented pasture.

Mark and Kristin Kimball, owners of Essex Farm on the New York side of Lake Champlain, are also at their kitchen table at five o'clock this morning, scheduling the day's activities. Kristin will prepare breakfast for the family and Mark will get the two girls scrubbed and dressed. Mark will then head to the barn and hitch two Belgian draft horses to an ancient cultivator, which after breakfast he will drive to the south fields to prepare the soil for planting.

These farmers, and others you will meet in this book, are smart and well educated (often with Ivy League diplomas and advanced degrees). However, some of their methods — the replacement of tractors with draft animals, for instance — leave "conventional" farmers scratching their heads. These farmers are part of an emerging movement in agriculture that is changing the way people grow and market food, and the way consumers relate to their food and to the people who produce it. I predict that as you get to know these farmers and begin to understand why they do things the way they do, you will come to appreciate their passion and to value their products. You may even join the movement, if you have not done so already.

The emergent agriculture is grounded in the philosophies of sustainability, local production, and the values of small-scale, family farming. The emergent agriculture values the crafts of the land and engenders not simply the ability to produce food (as if producing food were simple), but the ability to produce safe, nutritious food for dozens, hundreds, even thousands of consumers, and to do so in a manner that does not deplete the earth, that is profitable for the producer, and humane to the organisms in one's care. In short, the emergent agriculture represents an alternative to what is increasingly recognized as an unsustainable industrial system.

As an ecologist, my job is to observe and explain the interactions that occur between living organisms and their environment. I've been at it for nearly 40 years. When I began farming about a decade ago, two things were immediately apparent. First, is that the farm is an ecosystem, structured by the farmer and functioning as a compromise with wild nature. Second, I discovered that I was no longer an observer. I was a stakeholder.

PAM KLEPPEL

I began to realize that my interpretation of the observations made from inside the system were not always the same as those made from the outside. I quickly realized that this perspective would improve my professional understanding of ecology. What I didn't realize initially was how deeply it would inform every aspect of my life.

From what I've learned about the farm ecosystem and the business of farming, I predict that increasing numbers of farmers will abandon industrial food production and commodities-based marketing, preferring the more appreciative, humanizing, and often more lucrative alternatives that are emergent. Clearly, I am no fan of industrial agriculture, but I don't condemn the farmers who participate in it. If you attended an agricultural college in the past 60 years, that is how you were taught to farm. It is probably how your parents farmed. It was, for most farmers, the only game in town. My distain is for the individuals and institutions — often represented by multinational corporations — that abuse farmers by creating bottlenecks in production and distribution chains, by privatizing elements of the food system (such as plant genomes) that have always been in the public domain and belong there, by ignoring the value of craftsmanship, and by turning crafters into anonymous units within a black box of proprietary food production. I deplore those who promote systems of farming that degrade the earth, jeopardizing the security of the American food supply and the health of the consumer. And I cannot tolerate the inhumane treatment of livestock by an industry that considers such treatment an unfortunate but inescapable part of routine protocol, and a government that accepts this.

The new system of agriculture is a long way from becoming the status quo. We are in the early stages of what is certain to be a slow, sometimes painful process of transformation from the industrial model to something better. If we are indeed witnessing a revolution in food production, we must seek to understand what is happening and the role we will play in it. The transformation must be considered in context, as one of a series of revolutions that have occurred over the ten thousand-year history of agriculture. The most recent of these began in the 18th century as part of the industrial revolution. Even today, it continues to evolve. But the revolution that gave us the industrial model of agriculture has run its course and has turned negative. And a new paradigm is taking its place.

The cornerstone of the industrial revolution was technological innovation. Farmers embraced science and technology as the means by which the efficiency of production and the quality of the product could be improved. Industrial agriculture brought us selective breeding before Darwin and Mendel were even born. It brought us new ploughs and tilling techniques, and numerous other tools and protocols. Production was ramped up to support the growing urban population needed for large-scale manufacturing.

In the early days of the transition to industrial agriculture, farmers had access to few external inputs. Fertilizers consisted of composts and manures produced on the farm. By the early 20th century, however, fertilizers produced by chemists working in the emerging agribusiness sector began making their way onto farms, replacing what livestock produced in the barn and fields. Synthetic fertilizers were developed on the "reductionist" premise that production could be enhanced simply by increasing the total amounts of the major plant nutrients — nitrogen, phosphorus, and potassium — in the soil. The chemists, however, lacked the deep appreciation for the soil that is required by those whose job it is to produce large quantities of food for extended periods of time. While crop yields from synthetic fertilizers were quite good initially, they eventually wore out the soil. Yields declined and susceptibility to disease and pests increased.

Soil, the industry argued, consists of a mixture of minerals, salts, and organic compounds that form a non-living, physical substrate within which plants can grow. Bacteria, fungi, viruses and other living members of the soil community were considered independent of the medium. It follows, the "reductionist" argument goes, that a knowledge of soil geochemistry was all that was needed to produce a better substrate than the original. The counterpoint to this thinking was the older theory of "humus" — that soil consists of a complex and somewhat mysterious fusion of microbial and geochemical components that together form a living medium which, if disassembled, fails to be soil. Justus von Liebig (1803–1873), a leader in reductionist thinking, claimed that the theory of humus would be debunked as the individual components of soil became more fully understood. He was wrong.

Long before the scientists who study biological complexity taught us that systems like soil cannot be understood by deconstructing them, Sir Albert Howard (1873–1947), the father of modern organic agriculture, demonstrated the inherent failure of agricultural reductionism. Sir Albert showed us what composts — essentially humus facsimiles — could do.[3] He challenged the industrial model of soil fertility. Sir Albert, himself a Cambridge-trained mycologist, made a career of restoring the health of impoverished soils with composts, and generating amazing yields of unusually robust crops without the aid of industrial chemistry. Sir Albert appreciated the irreducible interactivity of the soil's living and non-living components. He focused on the symbioses forged between fungi living in the soil and the roots of the crops that farmers were producing. These symbioses, called "mycorrhizae", increased in crops grown in well composted soils. Plants containing mycorrhizae grew rapidly, and rarely became diseased or infested by pests. Sir Albert repeatedly challenged the industry to grow the same crops next to his — composts versus chemicals — and see who got the best yields. The industry never took him up on it.

Even so, organic and compost-based agriculture took a back seat to industrial chemistry. Organic farmers, students of Howard, and others, such as the anthroposophists who followed Rudolf Steiner and created the biodynamic theory of agriculture, were derided as "anti-progressive" mystics. They were considered unappreciative of the values of science and technology and unwilling to grasp the progress being made in modern agriculture. After World War II, petroleum became the driver of industrial agriculture. Cheap oil allowed corporations to produce and farmers to buy (with steep loans) enormously powerful machinery. These tractors and combines shortened the workday, or increased by orders of magnitude the amount of land that a single farmer could work. Agribusiness would cash in on the revolution by developing petroleum-based products that would double or even triple production. However, they would not improve the bottom line for most farmers or the quality of products for most consumers.

In the mid-1960s, corporate America continued tightening its grip on farmers and on the policy makers, who "regulated" both the farmer and the industry. Corporate agribusiness infiltrated government and politics in Washington, securing key positions in Congress, on Presidential Cabinets and in the Judiciary, while continuing to make the case that large-scale production was the future of farming. The application of technological innovation and industrial efficiency would ensure an abundance of food at a modest price for generations to come.

At the same time, Americans of both genders were working outside of the home. Expedience and convenience in the kitchen were crucial — the TV dinner and the fast-food restaurant became iconic of the modern American household. This was the age of NASA's Apollo Mission, when the United States would send men to the moon. This was the age of Tang, and of comfort in the belief that, like astronauts, we all would soon be taking our meals from a tube. This was the age of government scientists who produced "fish-protein concentrate" that, when sprinkled on rice, provided a nutritious meal to the nutritionally deprived people of the third world, and who failed to understand why people in third world cultures rejected the supplement because they did not consider it food.

By intensifying agriculture with petroleum-based fertilizers and pesticides, production increased dramatically and prices at the market fell. How food was produced, who produced it, and what was sacrificed to get it conveniently to the table were superfluous. Because the government was responsible for food safety and we had faith in the government, we could assume that our food was safe. That, and price, were what mattered.

By the end of the 1960s, increasingly large swatches of the economic safety net that kept farmers afloat when nature dealt them a bad hand were being cut out from under them. Exacerbating the situation, grain failures in the early 1970s led to a sharp increase in beef prices. The federal government responded by intensifying its efforts to control the production system, creating economies of scale that would drive food prices lower. Farmers were under constant pressure to specialize, mechanize, industrialize, and grow. Earl Butz, President Richard Nixon's Secretary of Agriculture, exemplified the sentiment of the Federal government with his infamous "Get big or get out!" slogan. Academic agricultural economists rationalized that approach and the best Land Grant colleges in the nation taught a generation of farmers how to create an operation that would ultimately be swallowed up by some sort of corporate production system. Small scale farmers actually began to believe that the future of agriculture did not include them — that losing their family's farm was progress. The cost of getting big was staggering and as farms collapsed, they were merged into mega-farms with thousands of acres under cultivation. Each farmer

G. Kleppel

specialized in one or two products. Farming was now about monocultures and the pesticides and other inputs that kept them going. Production was standardized, and centrally controlled by the industry. Success was based on yield. The farm was now a ten thousand acre assembly line, the farmer was a factory worker, and food was a commodity.

The industrial food system continues to evolve. Today's magic bullet is transgenics. The agent of control is the patenting of the genomes of food organisms by large multinational corporations. The target of the industrial food system is a naïve and complacent consumer, concerned only that something packaged and sold as "food" is abundant and cheap, and can be assumed to be safe. As a generalization we can say that this paradigm has been good for neither producers nor consumers. It has been most profitable, however, for those in the middle — agribusiness, chemical companies, wholesalers, and various other "middlemen." After more than 60 years in this system, it is clear that the industrial model of farming is not working for most farmers or for most consumers. (Heart disease, diabetes, cancer, and obesity are recognized consequences of the modern Western diet.) It works beautifully, however, for multinational corporations.

Despite the derision, the lack of funding for research by US Department of Agriculture (USDA) and Land Grant Colleges, and the labeling (by the industry) of organic and other "unconventional" farming practices as simultaneously "backwater" and "elitist", the proponents of what would come to be known as sustainable agriculture have persevered. These farmers, who work on razor-thin margins, understand that they won't get rich doing what they are doing. They continue, in part, because of the personal satisfaction they derive from practicing their crafts and from being appreciated as craftspeople. Chefs at the country's best restaurants understand the difference between industrial and artisanal products — the difference between locally grown produce and the stuff off a truck that just travelled 1,500 miles. Increasingly, the public is recognizing the difference as well. Growing consumer appreciation for agricultural craftsmanship and its contribution to the quality and safety of our food is driving the transformation in agriculture.

The tide began to turn more quickly in first decade of the 21st century with the publication of Eric Schlosser's *Fast Food Nation* (2002) and

Michael Pollan's *The Omnivore's Dilemma* (2006), which documented the corporate monopolization of the American food supply and which challenged the assumption that our food is safe. Consumers are now beginning to question the notion that food is a commodity, and to realize that the production of safe, nutritious food requires respect for the soil, for living organisms, for ecosystems, and for farmers. The emerging paradigm is already changing the way food is produced and marketed. The farm is once again being recognized as an ecosystem, and the farmer as an integral part of that ecosystem. The market is becoming a collaboration between farmers and consumers, where the availability of information about production processes enhances food security, and where the value placed on farming by consumers ensures sustainability.

Pam Kleppel

... we have to begin recognizing the fundamental incompatibility between industrial systems and natural systems, machines and creatures.[1]

— Wendell Berry

2

The Paradox of Agriculture

EACH YEAR, EARLY IN THE FALL SEMESTER, I ask the freshmen in my Sustainable Food Systems class — most of whom have lived in cities or suburbs their entire lives — to close their eyes and think about the words "American farm." I ask them to write down what they see in their minds' eyes. A typical response is "I see a barn, with one of those tall round things [a silo] next to it. There's a field of wheat blowing in the wind. There's a house. It is white, but it needs paint. It has a big front porch." Another student writes "I see a farmer on a tractor. There is a cow and some chickens and a little girl throwing food on the ground for the chickens. And there is a dog."

The American family farm is iconic. Even though the vast majority of us live in urban communities and know very little about farming, and only one percent of us actually engages in it, we cling to our roots as an agrarian people. In our national mythology farming is an American thing to do. It is a pure, honest way to make a living. Farmers are rugged individualists, simple but savvy folk, who mean what they say and who can be counted on when the chips are down.[2]

Few of us, if we close our eyes and think about the words "American farm," envision a fast, low flying crop duster spewing a cloud of toxic chemicals over a thousand acres of soy, or ten thousand cows in a filthy feedlot receiving steroid implants and being fattened on corn, or 250 thousand chickens with their beaks cut off, laying eggs that fall through the wire floor of their metal cages to a rack below.

I walk over to the computer in our "smart" classroom, log into Google, and type "agriculture water pollution." One of the first hits takes me to a US Environmental Protection Agency website that describes the non-point source pollution of America's surface waters. Non-point source pollution is runoff, derived from diffuse sources — lawns, parking lots, and farms. Runoff may contain any number of contaminants; including fertilizers, antibiotics, and pesticides. Non-point source pollution is among the most serious causes of contamination of American waters.

I read to the class: "Farms are the leading source of pollutants to the nation's lakes and rivers and the third most important cause of non-point source pollution of America's estuaries."[3] As home to some 2,500 hog farms that manage approximately 10 million animals, North Carolina has become the poster child for agricultural non-point source pollution. Many of these hog farms, factories really, are located in the watershed of the Neuse River. Wastes from hog farms in North Carolina are stored in putrid, sewage-filled sumps called "lagoons", the stench from which can be detected for miles. Alternatively, the wastes may be sprayed onto the land — sometimes near wetlands and drainage ditches where they can contaminate receiving waters. University of Texas scientists have found that, in addition to extremely high nutrient levels, the aerosols from these sprays contain large numbers of antibiotic resistant bacteria.[4] The lagoons are not watertight — they leak. At times they load as much sewage into the Neuse River and its tributaries as a city of 10 million people. Normally, bacteria that live in the river convert organic inputs to inorganic nutrients, such as nitrates and ammonia which are taken up as food by the algae and plants that live in the river. In return, the algae and plants photosynthesize, producing much of the oxygen (a byproduct of photosynthesis) that is needed for respiration by the animals that live in the river. At low levels, organic inputs help to keep the river ecosystem functioning efficiently. But at the enormous levels at which sewage is added to the river by the hog farms, the capacity of the river's bacteria to process the load is overwhelmed. The bacteria use oxygen faster than it can be replaced, leaving the water column severely oxygen-deprived, or hypoxic. When fish swim into these hypoxic waters they suffocate. In 2012, a hypoxic event in the Neuse River killed approximately 75 million fish in 33 days.

This is not atypical of the factory or industrial style of agriculture that produces most of America's food. It is the agriculture that over-produces government subsidized corn and soy in the Mississippi Basin, the agriculture that spews so much fertilizer onto its corn fields that half of the nutrients are never absorbed by the plants and are washed by the spring rains into the Mississippi and its tributaries. When that nutrient-rich broth reaches the naturally nutrient-poor waters of the Gulf of Mexico it supports intense algal blooms. When the rains stop and nutrient inputs cease, the algae no longer receive the nutrients they need to grow and they starve and die. Bacteria go to work breaking down the dead algae. Again, the process requires more oxygen than the system can supply and, again, the water column becomes hypoxic. Currently, a hypoxic zone about the size of the state of Rhode Island is a permanent feature of the northern Gulf of Mexico.

Next, I point my browser to "food recalls". In seconds the class is provided with more than two-and-a-half million hits. First hit: "US FDA Food Recalls." I open the site and at random pick a year — 2010. Page 1 contains about 30 recalls for the month of December. This is followed by an equal number for November. In all, there are eight pages of recalls for 2010. Many, perhaps even most, are recalls of processed foods, but ultimately the basic food components of processed foods come from farms. It is quite clear that the American food system is producing a lot of products that are not safe.

A convincing proof of that statement is found in the databases of the USDA, the agency responsible for securing meat production in the United States. In February 2008, 143 million pounds of ground beef processed by the Hallmark Meat Packing Plant in Chino, California were recalled by the USDA. The reason: some of the cattle waiting to enter the plant fell down and could not get up by themselves, so they were "helped" by plant personnel... with electric cattle prods and high pressure hoses. Still, some of the fallen cows would not rise. They were fork-lifted into the plant. That's illegal. Every animal that enters a US meat packing facility must walk in under its own power.

And how was this illicit activity discovered? Well, it wasn't by the on-site USDA inspectors, who are supposed to watch every cow as it enters the plant. Nor was it by the CEO of Hallmark, who was "shocked and appalled by these revelations." Rather, it was the Humane Society, whose undercover operations captured the abuse on video.[5] CBS News anchor Katie Couric, who reported the story, concluded her report by noting that "... it happens all the time."

In reality, this recall was not about the inhumane treatment of cows. The federal government would not recall 143 million pounds of hamburger because someone abused some cows. I Google "mad cow disease symptoms." Here's what comes up: "One of the most common symptoms of mad cow disease in cattle is difficulty standing. The animals may also find it difficult to walk and have ... problems ... [with] muscle coordination..." The reason why "downer cows" are not allowed into a slaughter house is that the inability to stand and walk is symptomatic of several neuropathies, including bovine spongiform encephalopathy — mad cow disease.

It's 6:30 on a Thursday morning in mid-November. I have just arrived at the small, USDA packing plant operated by Lowell Carson and his two sons, Lowell Jr. and Zack. Behind my truck I am towing a trailer full of lambs to be processed in the Carsons' packing plant. Zack, wearing a blood stained leather apron and a thick belt, from which hangs a sharpening steel, meets me at the gate to the holding pen that leads to the abattoir. He will personally perform the humane slaughter of my lambs. A short, husky man in a white lab coat and hardhat stands at the door next to the holding pen. On the hardhat are the letters "USDA." I open the trailer door. Some of the lambs cautiously walk to the edge of the trailer and stop. We wait. I whistle and call cheerfully, "C'mon boys!" Nobody moves. We wait. Finally, I climb into the trailer, walk calmly to the front end, behind the lambs, and move slowly toward them. The lambs bounce out. One of them walks over to Zack and sniffs. The inspector leans over, staring at him intently, and at the lamb. The young man's arms go up; he does not touch the animal. The lambs walk into the holding pen. Zack walks into the facility. The inspector follows. I imagine that the USDA inspectors at Hallmark do their best to ensure that the livestock that enter the plant are healthy and humanely treated, but how much can one see when 300–400 cattle are entering the plant every hour. Contrast that with the ability of the inspector to observe the condition of each of my lambs and the way they are handled at a packing plant that processes about 30 animals a day. Then ask yourself, whose ground meat would you rather eat? Or feed to your kids?

So there you have it — the paradox of agriculture. Our image of what we want farming to be and the reality of what it has become. The words unsustainable, dangerous, inhumane, dehumanizing, and toxic are barely adequate to describe that reality. Agriculture is an American Jekyll and

Hyde, and to the 99 percent of Americans who are not farmers, whose contact with agriculture is at the supermarket meat counter and produce section, there is no easy way to distinguish Jekyll from Hyde.

The industrial Hyde of agriculture tries to pass itself off as its alter ego. At the supermarket, the images with which we are presented are those depicting farming as we'd like it to be. There are icons on the walls that depict farmers on tractors — usually small, 1950s era machines, not the huge John Deeres required for modern industrial-scale production. We see cows on pasture and perhaps pictures of the "local" farmers who "produce food for us." There is always an organic produce section, next to shiny tomatoes from Mexico and unblemished greens, moistened by automated misters.

Words follow pictures, words like "cage free" and "free-range" and "naturally produced." What do these terms mean? Many have no official meaning. Those that do often reflect federal oversight of only a portion of the production process. For instance, the USDA organic certification of spinach regulates what can be done to the soil, and the plants while they are in the soil, but not the harvesting, packaging, or transportation of the product. There is little that is inherently sustainable about USDA organic spinach once it is harvested.

There are, of course, no pictures of feed lots full of feces-covered steers with digestive systems in various stages of self-destruction brought on by the force-feeding of grain, or crop dusters delivering their dose of glyphosate to the fields of genetically modified corn and canola, or hogs packed together so tightly that they can't even lie down. Of course, there are no pictures of cows being fork lifted into a slaughterhouse. But what do we really know about supermarket foods, other than the price? What information do we, as consumers, get from the industry?

The public is becoming increasingly interested in the food system. People are beginning to understand the limitations on what they can get from a label, and what they need to find out by knowing who produces their food, where it comes from, and how it gets to their plates. Consumers are embracing systems that take care of the land and protect the water, that value craftsmanship, and require that food production be an ethical process. This is the point at which the façade created by the industrial system begins to break down. It is the point at which the paradox of agriculture is resolved. It is the emergent alternative.

MARK SCHMIDT

The ultimate goal of farming is not the growing of crops, but the cultivation and perfection of human beings.

— Masanobu Fukuoka, *The One-Straw Revolution*

3

Farm Subsidies

I KNOW WHAT YOU'RE THINKING — that this chapter is going to be about the billions of dollars Congress serves up to agri-business to permit the sale of agricultural commodities at very low prices, or to pay farmers not to produce. In reality, relatively few farms receive most of these subsidies — about 11 percent of the farms in this country get 75 percent of the money. These are large farms, usually planted with thousands of acres of one or two of the "big four" crops — corn (grade 2, used mostly for livestock feed), cotton, wheat, and soy.[1] The subsidies allow corporate agri-businesses to create "bottlenecks" in processing or distribution systems. A bottleneck is a point within the food system where control of the product passes from a farmer to a corporation, under conditions dictated by the corporation. For instance, an Iowa farmer who grows corn for the industrial system produces an average of 187 bushels of corn per acre.[2] If this farmer is working 1,000 acres, he needs to dispose of nearly 200,000 bushels of corn once it is harvested. The farmer can't bag that corn. He can't sell it and ship it throughout the country. He is the grower. The job of distribution rests with the company that buys his corn. This is accomplished at enormous storage tanks called grain elevators, which are usually owned by large corporations.

The corporation sets the price at the grain elevator. The farmer can take that price or leave it. Of course the farmer, who has taken a loan to buy the corn seed, has to take it. Typically, the prices are set artificially low, thereby creating an artificially cheap product which sells well on the global market. For instance, in 2005, the cost of producing a bushel of corn was about $2.50. The price paid at the grain elevator was $1.46 per bushel. The idea has been that the federal government would pay the farmer a subsidy amounting to the difference between the price at the grain elevator and what is needed in order to be profitable. In reality, however, the subsidy has declined by about five percent over the past decade[3] — this for the guy in the huge, GPS-guided tractor who, embracing a perceived economy of scale, needs to process thousands of acres of monoculture just to break even. Unfortunately, the cost of operating those 350+ horse power tractors has not declined. Nor have the costs of the huge quantities of synthetic fertilizer and pesticide needed to sustain these monoculture crops. For instance, in 2009, Monsanto announced a price increase of 42 percent for its genetically modified corn seed.[4] Meanwhile, multinational corporations, which control most of the food and fiber production in America, pack the halls of Congress with lobbyists; as well as state and federal agencies, and courts, with former-executives. The Union of Concerned Scientists reported that in 2008, a single corporation spent $8.8 million on congressional lobbying activities.

Farm subsidies so reduce the cost of the things these agri-corporations sell that they can, for instance, sell corn in Mexico for less than the Mexicans can sell their own corn. We produce so much cheap (subsidized) corn in the United States that we have trouble figuring out what to do with it. We can't use all of it as corn, so we deconstruct it and use corn-based chemicals to make everything from lipstick to soft drinks to disposable diapers. The high fructose corn syrup in many of our foods and drinks is partially responsible for the nation's increasing rates of obesity, diabetes, and heart disease. Cheap, subsidized corn is turned into animal feed that can turn the digestive system of a cow into a culture vessel for the acidophilic *E. coli* O157:H7 bacterium that each year sickens tens of thousands of Americans and kills about 60 of us, according to the Center for Disease Control and Prevention in Atlanta.

But conventional farm subsidies are not what this chapter is really about. This chapter is primarily concerned with the ways in which farms subsidize us. First, most of the farmers who fall into this category of subsidizers own small to medium-sized farms, and they don't receive much in the way of government assistance. They are more likely to be involved in local economies and direct-markets than in global commodity markets.

What are these subsidies? They are not things we buy directly from farmers, such as vegetables or bread at a farmers' market, or things purchased indirectly from farmers, such as meat or produce from a supermarket. They are the things we get from farms that we don't pay for — such as clean air and water, biological diversity, and pastoral landscapes. Let's look a little deeper...

Each day New York City provides potable water for about eight million residents. New York City residents use more than a billion gallons of water a day. However, New York City's water doesn't come from New York City. Rather, it comes from a series of reservoirs — the so called New York City Watershed — about 125 miles upstate. Most of it never passes through a water treatment plant. It is cleaned as it passes through the wild and agricultural landscapes — forests, fields, and wetlands — in the watershed.

Several years ago the US Environmental Protection Agency noticed that the quality of New York City's drinking water was deteriorating. The cause was intensive suburban development in the watershed. The solution: build a water filtration plant at an estimated cost of $8–10 billion (excluding the recurring costs of staffing and maintenance). A group of economists and ecologists met at Columbia University to consider alternatives to this expensive path to potable water. They understood that the reason the quality of New York City's drinking water was failing was that the farms and forests that dominated the rural landscapes of the New York City Watershed were being replaced by suburban housing, strip malls, and miles of roads and parking lots.

When water runs into a grassy pasture or a through forest or a meadow, much of that water is absorbed into the soil and "processed" through a matrix of "sticky" organic particles, plant roots, fungal mycelia, and bacterial cells. These remove the minerals, nutrients, and even pollutants that the water may be carrying. Ultimately, the soil releases the purified water to a reservoir or stream. As long as this natural filtration system is intact, New York City's water gets cleaned free of charge. When farms and forests are replaced by suburbs and strip malls, with their enormous amounts of impervious asphalt, the filtration provided by the soil, and by the plants and microbes in it, is lost. Thomas Schueler of the non-profit Center for Watershed Protection in Silver Spring, Maryland estimates that significant deterioration in water quality occurs when as little as 10 percent of a watershed is covered by impervious surfaces.[5] The experts who met at Columbia University to consider alternatives to building that expensive filtration plant understood that if the farms and forests (and most of the forests in the region are on land owned by farmers) could be sustained, so would the quality of New York's drinking water — and at a cost estimated to be less than 10% of the cost of a filtration plant. As a result, New York City has been purchasing the development rights — the attributes in the deed to one's property that permit one to build on the land — of farmers and other land owners throughout the Hudson and Delaware Valleys of upstate New York. By keeping these farms as farms, New York City is ensuring the quality of its water supply for a long time to come. The farms are just doing what they do — subsidizing the cost of water purification for eight million people — every day. What's that worth, I wonder?

The word *biodiversity* refers to the variety of living things on Earth. It includes not just the hundred million or so species of plants, animals, fungi, bacteria, and ancient bacteria-like archeans that inhabit the planet along with us humans, but also the genetic variety found within any single species.

Compared to urban and suburban landscapes, farms are mini biodiversity hotspots. They are the home of an enormous array of livestock breeds, crop varieties, and forage plants. On my little farm we manage four breeds of sheep, as well as crosses between breeds. There are anywhere from two to six breeds of chickens on the farm every summer. In any spring, we'll plant at least two dozen species of berries, fruits, herbs, and vegetables in our gardens; and we may plant two or three varieties of each of our favorite crops. The domestic genetic richness that emerges from farms supplies a cornucopia of flavors, aromas, textures, and colors to our tables. Think about the variety of chili peppers (I've counted 61 so far), of apples (in 1900, 7,000 varieties of apples were cultivated in the United States), or potatoes (the Peruvians grow 1,200 varieties). There are 244 distinct breeds of domesticated sheep in the world today — all are the same species, *Ovis aries,* but each breed is genetically different, having been selectively bred to be perfectly suited to the place where it was developed, the place where the farmer who developed it lives (or lived). Breeds are constantly being "crossed" and re-crossed to optimize desirable characteristics — more meat, better wool and so forth. Some of these crosses may ultimately become distinct breeds.

Farms are also sanctuaries for wild biodiversity. Researchers in Britain report that land on small, organic farms tends to be "shared" with wildlife.[6] A rather non-intuitive example of the convergence of wild nature and domesticity was reported in a 20-year study of farming practices in Switzerland.[7] The authors observed that organic farms produced about 80 percent of the yield of most crops grown on conventional farms, with half the inputs, i.e. fertilizers and pesticides. They found that the diversity of microbes was significantly higher in organic soils than in the soils of conventional farms. Nobody put those microbes in the soil. They live there, along with fungi and all manner of invertebrates. The work they do in the soil, free of charge, increases its productivity and the health of crops, reducing the need for synthetic fertilizers and the plants' susceptibility to diseases and pests.

Sportsmen and women know that whether they seek deer or quail, pheasant or wild turkey, eventually their hunt will lead them to a farm. These large tracts of land, not-quite-wild but clearly not tamed by urbanity — full of fields and wetlands and woodlots, are the refuges, sometimes the last refuges, of wild biodiversity. Harvard biologists Robert MacArthur and Edward Wilson are credited with most clearly articulating the relationship between space and species.[8] They studied the birds on Pacific islands and found that as the area of an island increases, the number of species of birds also increases, but in an exponential manner. This species–area relationship has been applied to "islands" of open space formed by suburban fragmentation. As large landscapes are subdivided and fragmented by housing tracts and highways, shopping malls and parking lots, wild nature is lost at an exponential rate. Farms counteract this tendency and serve as refuges for wildlife. Even in areas where large, industrial farms are dominant, researchers have found that the presence of a park or preserve can stem the loss of biodiversity associated with intensive agriculture, and, sometimes even reverse it. By providing feeding and breeding habitat or just catching the "overflow" of wildlife from parks, farms protect wild biodiversity and contribute to the pleasure we humans derive from this "biodiversity subsidy".

Although the biodiversity at our tables and on our landscapes enriches our lives in myriad ways there is much more to this story. Biodiversity buffers us from disaster. In an era of rapid climate change, this may be the most important subsidy that farmland provides. In the 1840s, the people of Ireland grew potatoes. More accurately, they grew one kind of potato, and they grew it intensively. Under such intensive cultivation, it is only natural that a disease epidemic would eventually occur. The Irish potato blight wiped out the crop from 1845–1850, resulting in the starvation of approximately a million people, and a massive wave of emigration. During the period of famine, the population of Ireland declined by nearly 20 percent. By contrast, the farmers of Peru who, as I have said, grow over 1,000 varieties of potatoes, have never experienced a famine. Biodiversity is a buffer against such disaster.

Ecologists David Tilman and Stuart Pimm and their many colleagues at universities and research institutes around the world have put years of effort into studying the effects of biodiversity on ecosystems.[9] They have discovered exactly what the farmers of Peru discovered 1,000 years ago — that diversity creates stability. Whether farm or forest, wetland or desert, as the number of species and the genetic variety within species increases, biological communities become more resistant to disruptive changes in the environment, and more resilient — able to recover — after a disruption. As some of the last bastions of biodiversity in the urban fringes of

Pam Kleppel

America, farms subsidize our society's stability in an era of increasing social, economic, and environmental flux.

Every year, students from my classes visit our farm. As they debark from their cars and SUVs, I notice a change in them. Their posture is different, their shoulders seem to relax. They seem uncommonly cheerful and unguarded, even when the purpose of their visit is to perform a laboratory exercise in my pastures. They are genuinely curious about the solar panels and about the chickens (can you pet them?) and about the difference between hay and straw. When they pull carrots from the garden, rinse them off, and munch away — it's like they had never tasted a carrot before. The farm is a classroom, a laboratory, a refuge, and a reality check for anyone who visits. People come to watch my border collies herd sheep. They come to learn how to bake bread and to see wool being spun on a wheel.

It's not just my farm. Small family farms connect people to something deep, hard to describe, but real and understandable to just about everyone reading this sentence. There is a hunger — in fact, there is more than that — there is a *need* to connect to something simple, straightforward, and fundamentally embedded in the human spirit. The family farm provides that connection. While some are willing to pay for that connection, many get it free of charge, as a subsidy provided to the 99 percent of the American public who are not farmers, by the one percent who are.

In all of these ways farms subsidize us. They improve the quality of our environment and the quality of our lives. They stabilize us in times of great change. They underwrite our social psyche, our sense of who we are, and who we think we ought to be. All of this, and — oh, yes — farms produce our food as well.

Part II
Sustainability

MJ Jessen

Pam Kleppel

It's not a question of whether we can be sustainable,
but whether we choose to be.

— Gary Lawrence, Director of the Seattle Planning Department

4

Toward a Sustainable Agriculture

SUSTAINABILITY MEANS, LITERALLY, THE CAPACITY TO ENDURE. To many, the word connotes environmental protection, conservation, or some approach to development that is environmentally benign. A former director of the Columbia, South Carolina Planning Department claimed that sustainable development is a contradiction in terms. "How can you develop and not cut down trees?" His lack of understanding is not atypical. The mistake is in equating conservation with sustainability. They are related, to be sure, but they are not the same thing. So what is sustainability? What gives one the capacity to endure?

Actually, sustainability implies behavior — that people behave in ways that ensure the capacity to endure. The 1987 Brundtland Commission Report on Sustainable Development to the UN proposed the following definition: Sustainability "... means that we meet the needs of the present without compromising the ability of future generations to meet their own needs."[1] The commission's report provided guidance to governments on how to promote economic development without sacrificing the long term integrity of their natural resources. Emergent from the Brundtland Report is a set of principles that focus on three fundamental elements of durable societies. They are: environmental stewardship, economic viability, and ethical behavior. These elements, when taken together — and only when taken together — are prescriptive of the long term survival and wise use of the ecosystems and social systems upon which our species depends. Sustainability is a set of behaviors, a way of thinking about how the present affects the future that leads simultaneously to environmentally, economically, and socially desirable outcomes. While it is questionable whether governments have paid attention to the Brundtland Report, it is apparent that many others — from individuals to corporations — have.

The word sustainability is widely used today, so much so that some feel it has become a buzz-word, a marketing strategy, a form of green washing. As a supporter of the PBS News Hour, for instance, the Monsanto Corporation proclaims its dedication to sustainable agriculture. Make of that what you will. My take is that corporations such as Monsanto would not lay claim to sustainable practices if they felt that sustainability was unimportant or if they felt that the market was uninterested in the concept. The drive to survive, to endure, is in our DNA. In one way or another, it is in the DNA of all species. The difference between us and other species is that we can predict the end game. We have the capacity to understand where our lifestyles lead. The question is whether or not we will pay attention.

Nowhere are there better examples of sustainability, *and unsustainability*, than in agriculture. By examining sustainability through the lens of agriculture one gains a deeper appreciation of its meaning and of its critical importance.

A few years ago my friend Jim Hayes, from Sap Bush Hollow Farm, was talking to one of my classes about his concept of farming. "The industry," he said, "has a single bottom line — profitability. I have three bottom lines: Profitability, for sure. My farm won't survive if I'm not profitable. My second bottom line is to take care of the ecosystems — the streams, the grass, the soil — that provide the resources I need to farm the land. And my third bottom line is to treat my livestock ethically and with respect. When you, as a consumer, have to pay a little more to buy my meat, it's because you are accepting my triple bottom line." That phrase, the triple bottom line, has become synonymous with sustainable agriculture.

Farming in and of itself is not inherently sustainable. Throughout history, dozens of cultures, from the Mesopotamians to the Mayans, have collapsed as a result of their agricultural practices. In modern times, the unviability of so many small and medium-sized farms and the current insecurity of our nation's food supply are testimonials to the unsustainable nature of current agricultural practices. When thousands of people are sickened by hamburgers and spinach, when livestock are routinely abused, when the genomes of our most important crops are removed from the public domain by a few multinational corporations, agriculture clearly has no inherent claim to sustainability.

While sustainability arises from the conscious decisions of the farmer to adopt the principles engendered by the triple bottom line, there is no prescription for how to achieve it. Gaining organic certification, pasturing poultry, putting up solar panels — these are all worthwhile, but they are not definitive of sustainability. Above all else, the process must start and end with ethics. Sustainability does not arise from stewardship of the land but from an understanding of one's ethical responsibility to *be* a steward of the land. When our first commitment is to ethical behavior we realize a deeper reason for our actions. Installing solar panels, rotating livestock and resting grazed pastures, replacing tractors with draft horses, adopting organic practices, saving seeds, using low and no-till planting techniques, and myriad other procedures, all speak to our recognized responsibilities toward the earth, its systems, and its residents.

The decision to seek organic certification does not make an operation more sustainable unless the farmer understands the underlying ethic of organic farming. Thousands of acres of industrially-managed vegetable monocultures owned by enormous corporations are officially designated "USDA Organic." The very fact that they are monocultures, however, violates a fundamental ecological principle — that diversity creates stability. The fossil fuel intensive harvesting, packaging, and distribution of the crop suggests a commitment to a system that is not sustainable, regardless

Pam Kleppel

of the label. Such operations are focused on the positive financial impact that the USDA Organic designation brings to the bottom line while ignoring the environmental responsibilities engendered by that label and, therefore, also ignoring the ethical values that must be brought to any sustainable process. As a result, many USDA Certified Organic farms fail to meet any meaningful criterion for sustainability, and they may actually threaten the long term viability of the organic farming movement.

Sustainability, therefore, is determined not by some government designation or label, but by one's behavior, and the "footprint" that behavior leaves behind. When corporations limit our access to the process of producing food, and when they control the information that we get about that process, it is easy to attach labels — organic, free-range, pasture-raised, all-natural — and appear to have a legitimate claim to the "high ground." Nature, however, distinguishes between labels and content. Content, not labels, determines our durability. Nature responds to actions, not promotions. If we're abusing the land, if we're abusing our livestock, if we do not respect farm workers, and if we treat the food they produce like commodities, we will not endure for long, regardless of what we call it.

Sustainability is a journey, not a destination. We can always do better. How hard we work at meeting sustainability goals is as important as achieving those goals. I once mentioned to Mark Kimball of Essex Farm that his approach to farming was the most sustainable I have seen. Mark's response was that he thought Essex Farm to be about five percent sustainable; his wife, Kristin, felt they were only one percent there. But they continue to strive to do better. Sustainability is defined by our adherence to the ethic of farming and our striving to do better. That striving to do better, I believe, is the universal thread in the fabric of sustainability.

An old African proverb, related to us by Wangari Maathai (1940-2011), who won the 2004 Nobel Peace Prize for what the committee cited as "her contributions to sustainable development, democracy and peace," drives home the point of my essay. Maathai was the first Kenyan woman to earn a PhD and to serve on a university faculty in her country. She understood the relationship between environmental stewardship and social equity, and she turned those concepts into realities. Maathai founded the Green Belt Movement, which paid women to plant trees to reforest the land. The Green Belt Movement has planted more than 30 million trees across Africa. As they planted trees, the women of Kenya gained dignity, and as they gained dignity they demanded and eventually were granted their civil rights.

Wangari Maathai told a story about a hummingbird and a forest fire.[2] While all the other animals of the forest watched the fire consume their home, a hummingbird flew to the river and filled its tiny beak with water, and then rushing to the forest, sprayed the water on the fire. The other animals chided, "You can't put out that fire! You're just a tiny hummingbird!" "But", said the hummingbird, "I have to do something. And I'm doing the best that I can." If the other animals were inspired by the hummingbird, they would save the forest. If not, their home would surely burn. Ultimately, Wangari Maathai never saw herself as a Nobel laureate, or the leader of a movement that would plant 30 million trees and achieve civil rights for women in the bargain. She never expected to "save the forest". But she knew that she had to do something. And she did the best that she could. We need to be inspired by Wangari Maathai and by her story of the hummingbird. And if we are, we will understand sustainability, and perhaps we will save our home.

PAM KLEPPEL

I dislike the thought that some animal has been made miserable to feed me. If I am going to eat meat, I want it to be from an animal that has lived a pleasant, uncrowded life outdoors, on bountiful pasture, with good water nearby and trees for shade.

— Wendell Berry, *What are People For*

5

Sustainable Meat – A Contradiction in Terms?

AMONG THE MOST COMPLEX AND CONTENTIOUS ISSUES surrounding the debate about what is sustainable and what is not, is the question of meat. Can it be produced sustainably? Should we eat it? As an overarching premise to guide this discussion, let me start with a simple statement: *It's about the process.* The product we get from any agricultural system is a function of how it is produced. There seems to be a growing consensus that eating vegetables is a sustainable process and eating meat is not. I believe that this is incorrect. I will argue first, that no food — plant or animal — is in and of itself sustainable. Its connection to sustainability is a function of how it is produced, i.e. the degree to which the process is sustainable. My concern is that certain ideologues who abhor the consumption of meat, for any number of reasons — some of which are quite valid — have foisted on the public a suite of myths about meat production that is fundamentally wrong in its logic and does not move the debate forward in any positive way. I'll address some of these myths with data later in this chapter. There are many good reasons why people should reduce and even eliminate meat from their diets. There is no need to make inaccurate arguments to promote that position.

Humans eat meat because we are omnivores. People in most cultures eat other animals — from insects to antelopes. We are designed to have some meat in our diets. However, a naked human running through the grass with a stick, 20,000 years ago, was not a great hunter. Even when hunting in groups, our ancestors rarely returned home with copious quantities of game. As such, meat became a special food — a communal food. Religious ritual frequently demanded a gift of meat to the deity.

And in many societies, the violence associated with harvesting meat is redressed by some form of ethical or spiritual capitulation to the prey, the divinities, or both. To have meat — and a lot of it — was, and still is, a sign of social status. But, even today, most humans rarely get much meat on their plates. In 2012, global per capita meat consumption was less than four ounces per person per day.[1]

Americans eat a lot more meat than that. In fact, at an estimated 276 pounds per person per year, we eat, on average, more meat per capita than the people of any other nation on earth except Luxembourg.[2] We consume three times more meat each day than the global average. Not only do we consume comparatively enormous amounts of meat, but, thanks to our industrial production system (which helps to make it abundant and cheap) much of the meat that Americans consume is not safe. For starters, meat produced from animals that spend the last four to six months of their lives in feedlots, where they are fattened on corn, are high in potentially artery-clogging, low density lipids (LDLs) and depleted of the more favorable high density lipids (HDLs).[3,4] Tissue levels of omega-3 fatty acids, the precursors of HDLs, decline exponentially over time when cattle in the industrial meat production system are moved from pasture to a feedlot.[5] Ultimately, our culture of meat production and consumption is unethical to our livestock and unhealthy to ourselves. The public health and environmental consequences of such behaviors are well documented.

The processes of slaughtering livestock and butchering their carcasses are carefully regulated by state and federal governments (though the regulations are not always enforced). When "the kill" is performed properly, death comes quickly and with minimal pain. However, concern for the ethical treatment of farm animals at the moment of their death does not extend backward to their treatment while they are alive. In fact, the kill may be the most ethical part of the process. Economies of scale and the emphasis on maximizing yield allow the industry to justify sacrificing the humane treatment of livestock under the logic of efficiency.

It doesn't take much to figure out when animals are stressed, however. It's not rocket science. When hogs are packed so tightly in their pens that they are unable to lie down, they are not going to be happy. If the farmer is constantly administering antibiotics to his livestock, the animals are probably under pathogenic stress, and as the pathogens and parasites he seeks to control become resistant to those antibiotics, it is a certainty that the livestock will become stressed. If beak oblation is required to prevent the 250,000 "cage-free" birds in a hen house from killing each other, it should be obvious that the house is overcrowded and that the chickens are seriously stressed. It should not take the Humane Society to tell us that such practices are inhumane. But these practices are widely used in industrial meat production.[6] Although they are completely legal, they are neither ethical nor sustainable. They are permitted because the public has been removed from the process of meat production — excluded from the overcrowded pens and chicken coops, and the putrid, germ-infested swine barns — and told that if it is not done this way, the price of meat will rise precipitously. That is false!

There is an incredible counterpoint to industrial agriculture among farmers striving to achieve a personal standard of ethics in meat production. I have visited dozens of small and mid-sized farms that offer glimpses into their processes. Most of the owners of these farms believe that the public has the right, even the responsibility, to understand the process of meat production and to demand that that process be clean, safe, and humane from start to finish. Many offer farm tours, petting zoos, hay rides, and CSA gatherings. At our farm, Pam and I do sheep herding demonstrations with our border collies. We, like many others, believe that only when the public demands it will the industry clean up its act. And only when the public understands the process of meat production can it make such demands. The emergent agriculture provides a window through which the public can view the process. Transparency helps to create and maintain an ethical foundation for agriculture.

Although I produce meat for the market, no one owes me an explanation if they choose not to consume meat. Several of my students and close friends are vegetarians. Some are vegans. One of my students and two of my friends who are vegetarians do, in fact, eat meat when they visit my farm. They believe that the meat we serve is clean, safe, and ethically produced. My friends and students understand that vegetarianism, per se, is not a synonym for sustainability. The ethical basis of vegetarianism is as easily questioned as that of eating meat. Tofu, the mashed soy-bean curd from which many meat substitutes are made, is often produced by a totally industrial process. Ninety percent of the soy-beans grown in the United States are harvested from genetically modified plants. These are grown in enormous monocultures, sprayed with pesticides, often over-fertilized (usually with phosphorus), and sold in the commodity markets. Vegetable production is inherently neither more nor less sustainable than meat production. Nor are vegetables necessarily safer than meat. For example, in 2008, the anaerobic acidophilic (acid loving) bacterium *E. coli* O157:H7, which proliferates in the digestive tracts of feedlot cattle force fed a diet of corn (which causes the normally pH-neutral rumen to become acidic) somehow ended up on spinach from the largest organic farm in the country. I have yet to find a satisfactory explanation for how *E. coli* got onto that spinach.

The point is that, whether one is a vegetarian or an omnivore, the assumption that the food supply is safe is generally not a good one. Industrial farming, whether organic or conventional, creates dangerous products. Food safety is not determined by a label (e.g., organic, naturally grown), but by the actual production *process.* Only when one understands the process can one make informed decisions about food safety. That understanding is often not conveyed in a label. It is conveyed by a producer willing to make the effort to provide information to the consumer.

Over the past few years, there has been a noticeable increase in efforts by committed vegetarians to convince people to give up meat. While I completely support efforts to help the public understand the importance of increasing our intake of plants and decreasing our intake of meat, I am concerned that too much misinformation is entering the conversation and that the debate is being oversimplified, and possibly slanted. I am concerned that many of the arguments being made, while usually well intentioned and appearing logical on the surface, are based more on ideology than data. For instance, on April 12, 2012, *The New York Times* published an op-ed piece by James E. McWilliams, a history professor at Texas State University, San Marcos, entitled "The Myth of Sustainable Meat." In that piece McWilliams argues that, despite the sincerity of farmers seeking to do the right thing, the negative environmental impacts of domestic animal production are unavoidable. He suggests that livestock will always be treated unethically and that if we wanted to ramp up pasture-based meat production to where it could support the majority of consumers, we would run out of land. Furthermore, he claimed, farmers would cut corners to get the edge on their competition. Finally, McWilliams speaks to the fact that nutrients, in the form of grains and manures, are unnaturally moved about the biosphere to support the so-called "sustainable production of meat."

McWilliams takes particular aim at Joel Salatin, a third generation farmer and owner of Polyface Farm, in Virginia's Shenandoah Valley, claiming that Salatin, who feeds his chickens grain produced "off-farm," is violating the natural rules of nutrient transport.[7] This, of course, is not the case. Nutrients in nature are highly mobile — they have to be, because nutrients are the stuff from which living organisms are made, and as organisms live, move about, die, and decay, nutrients are transported throughout the biogeochemical landscape. Water, air, animals, plants, bacteria, and fungi move enormous amounts of nutrients around the globe every year.

McWilliams also suggests that Salatin's use of chicken feed produced off-farm obviates his efforts to be a good steward of the land and an ethical manager of his flock. But virtually all farming involves taking nutrients from one place and putting them someplace else. The good steward ensures that the nutrients one puts on the land don't end up someplace where they can do harm. Thus, the techniques of cover cropping, intensive rotational and "high grass" grazing (some of which Salatin pioneered) tend to retain nutrients in the soil and to prevent them from leaching into streams. Vegetable production also depends on inputs, on nutrients moved from one place to another, as does all plant life in both wild and cultivated systems. Finally, it needs to be said that, among those of us practicing "mythological" meat production, Salatin is considered the very icon of sustainability. He vigorously pursues environmental management and business practices that have made Polyface Farm a model of ecological functionality and economic viability in agriculture.

Salatin responded to McWilliams' attacks by pointing out that many of the professor's comments evidence a poor understanding of modern, pasture-based agriculture and are simply inaccurate. That was also evident in an op-ed about pork production (in 2009) in which McWilliams argued that pastured pork was less safe than the industrial alternative. McWilliams' arguments were in that case rebuffed by Dr. Shannon Hayes, a third generation farmer and well published author in her own right. As an academic who participates in agriculture, I can add that it is one thing to sit in an ivy tower and know the *theories* of food production. It is quite another to be a farmer and know the *realities* of food production.

Simplistic arguments and misinformation do little to improve the quality of our food, the sustainability of its production, or the character of the debate. A good example is the claim that there just isn't enough land available to pasture the nearly 100 million cattle in the "national herd." The truth is that the amount of land required to finish a cow on grass varies with so many factors that it's not even calculable, except on a very local or case-by-case basis. Pasture quality varies enormously throughout the country, even within states, counties, and townships. So do cattle breeds (not to mention the crosses between breeds) and management approaches. If cattle are just put out to pasture and left there, it is likely that they will overgraze, no matter how good the grass is. Under that management scenario, there probably isn't enough land to pasture the national herd. On the other hand, if the cattle are densely stocked and rotated frequently (mimicking the way wild ungulates aggregate and move naturally)[8] and the pastures from which the stock are removed are adequately rested, a relatively small amount of land can often support a relatively large number of cattle, even in arid climates.[9] The idea that every cow requires 20 acres of land (per McWilliams) is simply inadequate to account for the complex and convergent influences of landscape attributes, breed attributes, and management approaches. Published pasture requirements for cattle range from 0.5 to 20 acres per cow. I rotate about 20 sheep (about 1.5 tons) through quarter acre paddocks on six acres of moderate quality pasture throughout the season (usually April or May to November). My sheep are healthy. Just doing the math (and ignoring the rotational process), I'm pasturing about a quarter ton (500 lb.) of sheep per acre. If we ramp that up to a 1,400 lb. steer, we would need about 2.8 acres to raise that animal — about fourteen percent of McWilliams' suggested spatial requirement. Jim Hayes, in nearby Schoharie County, estimates that his cattle and sheep graze between 1.5 and 3.0 percent of their body weight per day and that his pastures produce about 1.5 to 2 tons of vegetation (expressed as dry matter) per acre. Depending on whether we have a seven month (214 days) or a ten month (304 days) grazing season, the estimated land required for a 1400 lb. ungulate ranges from

2.1 to 2.9 acres, fairly close to my estimate and a long way from 20 acres. But again, what works on my pastures may not work across the country, my state, or even my town. To add a little more complexity, several other species, say sheep and chickens, can use the same land that cattle graze — the method is called inter-grazing — and the results (for grass and soil quality) have been quite good. Increasingly, farmers are inter-grazing sheep or goats with cattle. The goats browse on the woody saplings trying to invade the pasture, leaving more room for grass and herbs, which will feed the cows. I inter-graze sheep and chickens. The chickens follow the sheep, foraging for parasites and scratching the nutrient-rich dung into the soil, helping to improve fertility.

In addition to the approximately 409 million acres of land currently being used to graze cattle, a third of the 58 million acres now used to grow grade 2 corn principally for livestock feed[10] could be re-purposed for grazing. If the nation's meat producers were to switch from the industrial, feedlot-based meat production system to pasture-based systems, the need for grade 2 feed corn would decline significantly. The land being used for corn could be turned back to pasture (perhaps alternating with crop production) which would reduce the need for synthetic fertilizers (since animals provide natural fertilizer free of charge). If only one-third of the land being used for grade 2 corn production were turned back into pasture, 19 million acres of pasture would become available. Adding this 19 million acres to the 409 million currently in use as pasture, gives us 428 million acres available for meat production.[11] Is that enough? We really won't know until we take into account the capacity of every acre of land to produce grass, until we know what kinds of cattle farmers and ranchers are raising, and until we decide how those animals are going to be managed. But by simply doing the math — dividing the amount of pasture by the size of the herd — we find that about 4.3 acres are available for every cow in America. That's about a quarter of what McWilliams says a cow needs, nearly 10 times more than the literature-minimum, and about 1½ times more than what I estimated would be needed, based on my experience with my land and grazing techniques. In conclusion, while grass-based agriculture is not likely to become the norm for American meat production any time soon, the claim that the potential for pasture-based meat production is space limited seems simplistic, at best.

While we're debunking myths, let's take a look at a couple of others. One of the most common arguments for quitting meat is that cows require enormous amounts of water.

It is true that cows drink a lot of water. A single cow can drink more than 25 gallons in a day.[12] To put that into context, however, we humans use between 140 and 170 gallons of water per person per day, and there are more than three times as many of us as there are cows. So the nearly 100 million cows in the United States consume about 2.5 billion gallons of water daily.[13] That's 0.5 percent of the 500 billion gallons (which includes personal, industrial and other uses) that Americans use each day, according to the US Geological Service. More importantly, nearly half of the water consumed by a cow doesn't stay there. If you've ever seen a cow pee, you have no doubt about this! An average cow has a urine volume of about five gallons per day and a fecal water content of about six gallons per day.[14] Depending on the temperature, humidity, and other factors, another half-gallon or so is respired — making a total water loss of about 11.5 gallons per animal per day, or about 1.15 billion gallons per day for the American herd. As a result, about 1.35 billion gallons of water are incorporated into our cattle each day. That's only about one-third of one percent of the national daily freshwater consumption. We're not going to run out of water because of cows.

We sometimes forget that most of the water consumed by animals doesn't stay there. A pound of beef (which is between 56 and 81percent water) and a pound of soy (61 to 79 percent water) have similar water contents. They are not much different from us (57–79 percent). So what happens to all of the water that a cow excretes? That depends on where it's excreted. In a feedlot, it may be stored in an impoundment or diluted and allowed to run off or evaporate. It may be managed to prevent it from transporting contaminants. On well-managed pasture, however, water as urine, and as a component of feces, is a vehicle for delivering organic nutrients to the bacteria and fungi that live in the soil. These organisms help decompose the organic nutrients into the inorganic constituents that vascular plants need to grow. Researchers studying the nitrogen cycle (nitrogen is important because it is a component of protein) in moist soils at temperate latitudes have found that dung and urine from grazing ungulates (hoofed mammals) stimulate microbial activity, which tends to hold the nitrogen in the soil rather than allowing it to leach out, making it available to plants and enhancing their growth.[15] Thus, on well-managed pasture, the aqueous portion of all that urine and fecal material is "cleaned" by the organisms living in and on the soil, as it moves, much of the time below ground, through the hydrosphere. Bottom line — the fact that cows drink water is probably not a sign of imminent disaster.

A final illustration of oversimplification in the debate about the sustainability of meat production is that the methane ruminants emit contributes significantly to climate change. There is no doubt that methane, which is about 20 times more effective than carbon dioxide at keeping heat from leaving the planet, is a potent greenhouse gas. Cows, sheep, and goats exhale methane as a result of the fermentation that occurs in their rumens (i.e. digestive tracts). Enteric fermentation contributes about 1.5 billion metric tons of methane to the atmosphere annually. That amounts to 23 percent of all methane emissions (6.7 billion metric tons per year) and two percent of the annual greenhouse gas emissions from the United States.[16] Enteric emissions from ruminants are nothing new — the herds of bison and other wild ungulates that once roamed the American Great Plains were emitting methane long before humans arrived on the scene. But, like the "cows drink too much water" argument, it seems that the ruminant-methane story is intended to convince us that the rumen of a cow is an ecologically dangerous organ. In truth, the issue is complex, and if we are going to make decisions about whether or not to quit meat because cows belch we should face the matter with facts.

When a cow eats grass, un-mowed and un-amended by petroleum-based fertilizers, that grass represents contemporary carbon — the carbon budgeted to creating that grass is derived solely from carbon dioxide in the air via photosynthesis. When processed in the rumen, methane is produced as a by-product, but no new carbon is added to the global carbon cycle. The total amount of carbon "in play" in the upper crust of the planet and in the atmosphere doesn't change. When corn grown with fossil fuel-based fertilizer is fed to cows and the cows exhale methane, the amount of carbon in the system increases by an amount proportional to the amount of "fossil" carbon, i.e. fossil fuel based fertilizer, used to produce the corn. Long-buried carbon has now entered the contemporary carbon cycle as a potent greenhouse gas.

The second part of the issue has to do with photosynthesis — the conversion of carbon dioxide to carbohydrates by green plants. Essentially, photosynthesis is the way plants make food. The photosynthetic activity of the plants in a pasture results in the sequestration, i.e. removal, of atmospheric carbon dioxide. When ruminants, such as cows, sheep, and goats graze on pasture, even though they are putting carbon (methane) into the atmosphere as a function of their respiration, carbon is simultaneously being removed from the atmosphere by the plants — grasses and weeds. When plants are actively growing, the amount of carbon removed from the atmosphere by photosynthesis far exceeds the amount emitted into the atmosphere by the respiration of the grazing animals. As photosynthesis slows in the fall, respiratory carbon output begins to catch up with photosynthetic carbon removal. Livestock, now eating hay, continue to emit methane. That has to be accounted for in any determination of a farm's "carbon budget."

A few years ago, I calculated an annual carbon budget for my farm. I included in the budget all of the animals, plants, and machinery used in my operation. I also included the pastures and the plants in the hedgerows, but not the lawns and gardens around the main house. I did not include in my analysis the carbon used to produce such things as chicken feed, which is produced off my farm and transported in. I did include the amount of carbon emitted during my trips to and from the feed store. Much of the data, and many of the constants and conversion factors in my analysis were not my own measurements; they were "mined" from the literature and from existing data bases. So my estimates are rough. Yet the results are clear — our farm was carbon-negative by a couple of hundred tons per year. That is, far more CO_2 equivalents (the effective amount of CO_2 represented by a greenhouse gas — for example, one gram of methane equals 20 CO_2 equivalents) were sequestered by the plants in our pastures and hedgerows than were produced by our animals and our greenhouse gas-emitting operations.

My results were consistent with those of Dr. Rita Schenk of the Institute of Environmental Research and Education in Vashon, Washington. She compared the number of CO_2 equivalents produced by cattle in confined animal feeding operations (CAFOs, i.e. feedlots) and on pasture. In CAFOs cattle add, on average, nearly 4,000 CO_2 equivalents to the atmosphere during their lives. On pasture, about 200 more CO_2 equivalents are sequestered by photosynthesizing plants than are produced by the cattle. Even allowing for the impact of CAFOs on the carbon budget, scientists at the US EPA and at Cornell University agree that all agricultural activities in the US contribute only eight percent of the nation's total greenhouse gas emissions.[17] Furthermore, H. Alan Ramus and colleagues at the Louisiana Department of Renewable Resources and the University of Louisiana, Lafayette, found that intensive rotational grazing consistently resulted in lower methane emissions by pastured cattle than un-rotated cattle on pasture.[18] So not only does a grass-based grazing system remove more carbon from the atmosphere (storing it as living plant biomass) than a feedlot, but the strategy used to manage the cattle's grazing influences the amount of methane produced. The take home message is that the impact of livestock production on greenhouse gas emissions depends on how the animals are managed — *the process*!

The point of these exercises is not to demonstrate how well I can fit a cow into the earth's water and carbon cycles. Rather, it is to elevate the debate. There are lots of good reasons to be a vegetarian — from the poor quality of as much as 90 percent of the meat consumed in the US today, to the cost of getting a safe steak, and the ethics of domestication and modern production — but it is not because cows drink water and belch. The issue of sustainable meat production turns on *process,* not just product. Shall we eat meat, or any food produced in ways that degrade our natural and cultural resources? The same questions apply to all foods.

That said, real challenges do lie in the path of pasture-based meat production and these should be discussed. First, management intensive and pasture-based grazing techniques, while well suited to small and moderate-sized flocks of sheep and herds of cattle and goats are more difficult to use with large numbers of animals. Second, pasture-based meat and dairy production has been set back by the loss of genes that traditionally allowed livestock to "finish" on grass.[19] Selective breeding "moves" the genetics of livestock toward specific outcomes, such as larger size, desirable wool colors, and greater milk volume. But when traits are of no interest, they may inadvertently be lost through disuse. We call this "genetic erosion." Over the past 70 years or so, the focus of breeding has been on factors other than improving metabolic efficiency on a grass diet because grain supplements could achieve the desired result — rapid growth in lambs for instance, or copious milk volumes in Holsteins — without grass. Livestock production has changed from a system focused on the grazer-grass relationship, to one based on the grazer-grain relationship. As feed-based farming has become dominant in agriculture, the now-unnecessary genes that once drove the grazer-grass system have become dormant or have been lost altogether.

A British shepherd who has been working sheep for 30 years once told me that it is impossible to finish a lamb on grass. "Just don't tell that to my lambs," I replied. "The only way you can do it is to rotate them every three days or so," he said. "Welcome to my world," I said. That's intensive rotational grazing. The livestock are fenced into a small portion of the pasture at high densities compared to those used in conventional grazing procedures. The fences are moved and the livestock are rotated onto new pasture at high frequencies — every three days, often sooner. It's really not the rotating that's important; it's the rest from grazing that the grass is getting. That rest period — fifteen to more than forty days, depending on environmental conditions and one's pasture management goals — allows the urine and dung left behind by the livestock to be processed by microbes in the soil. It allows the grass to grow unimpeded by large grazers. The whole process mimics the grazing behaviors of large, wild, herd forming ungulates, such as bison and wildebeest (that rarely overgraze their wild landscapes). These herbivores aggregate densely to discourage predation and move constantly to avoid contact with their wastes. Frequent rotations on pastures improve the quality of the grass. My students and I have found that intensive rotational grazing can be used to restore plant biodiversity in degraded landscapes that have been invaded by exotic plants. We've published several technical papers and reports describing our approach and findings.[20] We have found that when our sheep graze in these wild, biologically diverse landscapes they are healthier than sheep kept on pastures, which are essentially grass monocultures. But, as mentioned earlier, intensive rotational grazing is not well suited to large herds. Even with 20 head, divided into sub-flocks to separate ewes, rams, and ram lambs, the "intensive" part of intensive rotational grazing takes a toll on the farmer.

In the fall, as the growth rate of the grass in our pastures declines, we begin supplementing with hay. To run a grass-fed sheep operation, protein-rich second cut hay may be required. If you are not producing your own hay, purchasing the expensive second cut will surely affect the bottom line. The limits on flock or herd size, the cost of hay, and the relatively slow growth of the stock in the absence of grain-based feed, shave one's margins razor-thin in the labor-intensive process of pasture-based meat production. Given the overhead, the price of the meat simply has to be higher than that from steroid-laden feedlot cattle that have been stuffed with cheap grain and antibiotics for two to six months.

Selective breeding is beginning to recover the genetics needed for lambs and calves to grow and finish rapidly on grass. For the past 20 years, Jim Hayes has been developing a strain of Dorset sheep that can finish on grass. At Longfield Farm, we've been crossing metabolically-efficient Leicester Longwool rams with large Romney ewes to get a large lamb that grows quickly on grass — the jury is still out on our success, though my customers have been very happy with the meat, and the weight of our six-month old lambs averages about 85 percent of the weight of grain-fed six-month olds. Paul Van Amburgh at Dharma Lea Farm, in Sharon Springs, New York, is having success with grass-fed dairy cows that he imported from New Zealand. The craftsmen and women at Hawthorne Valley Farm, in Ghent, New York, are using the biodynamic model to manage a herd of about 60 Swiss Brown cows that produce high butterfat milk from which they produce cheeses that are sold at the retail store at the farm.

Among the most serious impediments to sustainable meat production is the shortage of small, independent processing and packing facilities. Because the number of meat packing plants in North America is limited, farmers and ranchers are often forced to transport their livestock long distances to be processed. The cost of transportation, of course, adds to the cost of the product. Moreover, by controlling access to meat packing facilities, the industry determines how much farmers and ranchers will be paid for their livestock. Eric Schlosser, author of *Fast Food Nation* (2002, Houghton Mifflin) describes how independent ranchers in Colorado have been forced to accept less for their livestock than the cost of production. This, says Schlosser, has led to an increase in the rate of suicides among ranchers.

For several reasons, problems associated with the shortage of packing plants have not gone unnoticed. It is clear that large, industrial packing plants are not producing a safe product. Oversight by USDA inspectors appears insufficient to achieve control over product quality, and inhumane treatment of livestock at these facilities is well documented.[21] Increasingly, the public is demanding greater oversight. Furthermore, there appears to be a resurgence of interest among young people in the crafts of meat processing and butchering. Small processing facilities and butcher shops are popping up throughout North America. Several states have developed their own meat inspection programs, relieving the pressure on the federal government (USDA) to provide inspectors, and allowing more small state-regulated facilities to be developed. Small-holder farms, selling meat locally, can use these state-regulated facilities in lieu of USDA facilities, which are only necessary if the meat is to be shipped across state lines.

Finally, there is the legitimate question about the cost of sustainably produced meat. Can the middle class really afford to eat pasture-raised meats? I believe they can. Michael Pollan, author of *The Omnivore's Dilemma,* points out that Americans place a higher priority on their consumer electronics than they do on their food. The reason, I believe, is commodification. When we believe that all food is the same (that's part of the definition of a commodity) then the only thing that matters is its price. Assuming all cell phones are not the same, people will pay more for a "better" cell phone. But if food is a commodity then, by definition, paying more doesn't get you more; so cheapest is best.

But Americans are beginning to realize that food is not a commodity. How our food is produced affects our health as well as the health of the animals being sacrificed for us, and the health of our ecosystems. When food has value, when consumers realize that the way food is produced affects its quality and safety, priorities change. We have a waiting list for our lamb. Fresh eggs always sell out at our farmers' market even though they're a dollar more per dozen than at the supermarket. And friends living on middle class salaries have switched from supermarket meat to grass-fed products from animals raised by farmers they know. By eating a little less meat they find they can afford to pay a little more for the superior, pasture-raised product.

Agriculture is constantly evolving. Farmers are always looking for better ways to produce food. When I meet with farmers from the industry, the focus is always on yield and profitability. When I meet with colleagues from the sustainable and organic agriculture communities, profitability is always embedded within the triple bottom line that includes concern for the land and for the ethics of farming. As this agriculture evolves farmers will return their livestock to the fields where they evolved, those animals will rediscover their grazing-genes, and consumers will realize they don't need to eat meat every single day to be happy. Perhaps then sustainably-produced, healthful, pasture-raised meat will be readily available to the majority of Americans.

Pam Kleppel

Environmentalists love the land; farmers marry it.

— Evan Eisenberg, 1996, *The Ecology of Eden*

6
Diversity in Agriculture

"BIODIVERSITY" REFERS TO THE VARIETY OF LIFE — the hundred million or so kinds of animals, plants, microbes, and fungi — that inhabit our planet. Biodiversity reflects the full range of form and function, of adaptation and evolutionary creativity among living species. But biodiversity is also reflected in the genetic breadth of an individual species and by the variety of habitats, ecosystems, and landscapes within which species have evolved and continue to evolve.

For the farmer, biodiversity is both domestic and wild, and a farm can be a place where biodiversity flourishes or from which it is excluded. The way farmers relate to biodiversity distinguishes between philosophies of farming. According to a British study, practitioners of large-scale industrial agriculture tend to constrain biodiversity to specific species and even specific clones of species.[1] Wildlife tends to be absent from (or at least unwelcome in) the farmscape. Small-scale, organic farmers tend to be more "welcoming" toward diversity. They tend to share the landscape with wildlife, and to diversify the production system. The distinction between the exclusion of nature and wildlife-friendly agriculture is usually not sharp, however. Most farmers make up their minds about biodiversity, particularly wildlife, on a species-by-species basis. Unlike the environmentalist, who usually sees native biodiversity as an unquestionable good, a farmer has a personal stake in the ecosystem and he evaluates the effects of every species, and even specific individuals of a species, in his territory. I'm fine with finding half a rabbit in front of the barn in the morning, or observing a coyote mousing in an unused meadow, but when the coyote starts testing my fences during lambing season, our relationship changes.

A farmer can appreciate the landscape and its ecology as much as an environmentalist. But unlike the environmentalist, the farmer's fortune is bound to that landscape, as much as is the coyote's. A farmer is not a participant observer, but a stakeholder. As a biologist, I applaud efforts to restore the gray wolf within its native North American range. As a farmer, quite honestly, the thought of half a dozen 60-pound wild dogs watching me drive my sheep to and from the pasture every day gives me the willies.

While predators have never been the farmer's friend and insect herbivores are usually unquestioned foes, there are those who understand that it is often what we do and the way we do it that determine whether a coyote or a potato luper is an annoyance or a plague, something to be tolerated or something to be suppressed or eradicated. And most farmers who farm close to the land (as distinguished from those whose farms are factories) understand the role that even enemies play in maintaining the ecosystem within which the farm exists. Coyotes will certainly take my chickens and lambs if given the chance. But coyotes also control rats, mice, voles and rabbits. Although we'll never be friends, in most cases we co-exist, as good ecological competitors do.

The relationship between farming and wild nature is complex. Though the industrial model argues for greater — perhaps complete — separation, the collapse of civilizations throughout history and the failure of the current industrial model teach lessons that we need to pay attention to. Simply put, we cannot separate ourselves from nature. Disasters occur when we try. Sir Albert Howard is known as the father of modern organic agriculture because he sought, with composts and manures, to simulate the soil's natural humus and restore the health of agricultural soils depleted by chemical fertilizers. He argued that pest or fungal infestations of one's crops simply indicate that something is not right with the way one is treating the soil. Howard proved repeatedly that when the farmer makes the necessary corrections to his practice, the infestation disappears. Removal of the pest without artificial interventions (e.g., chemical pesticides) becomes the measure of the farmer's *understanding* of agriculture — the soil, the crop — and his relationship to them.

Consider, for example, *Haemunchus contortus,* the barber pole worm. *Haemunchus* is a scourge among sheep flocks. Very few parasites can kill a lamb as quickly as an infestation of barber pole worms. The worm attaches to the gut wall, sucking blood and causing terminal anemia. The worms' eggs are egested in the feces and the emergent larvae can be re-ingested. One year, a friend of mine lost a third of his lambs to *Haemunchus*. Typically, farmers deal with the worm by administering anthelmintic medications that kill worms. Almost all of the worms are dead within a few days... almost. A few worms, however, are always genetically insensitive to the drug, and they reproduce. As a result, we have narrowed the genetic diversity of *Haemuncus contortus* to only those individuals insensitive to anthelmintics. So we switch to a new drug, which works for a while, but of course the problem recurs. And the cycle of drug → resistance → new drug → new resistance continues, each time reducing the genetic diversity of the pest, but at no time effectively resolving the problem.

Enter the chicken. *Haemunchus contortus* will never develop resistance to chickens. So on my farm, chickens form part of the front line in the fight against *Haemunchus contortus*. The success of our defense depends on maintaining the genetic diversity of the parasite. As the sheep are moved out of the barn each morning to begin their day in the pasture, our "layers" move in. The chickens are on the prowl for bugs and worms. They eat some of the large *Haemunchus* larvae in the barnyard which, along with other bugs and plants, contribute to the quality of their eggs. Over the past year we have also begun moving our layers around the parts of the pasture recently vacated by the sheep. In the absence of chickens, problems with *Haemunchus* (and with flies — chickens eat fly larvae too) tend to increase.

In addition to chickens, there are two other weapons in my arsenal. The first is rotation. I described in an earlier chapter how we divide our pasture into quarter acre paddocks. Our sheep are moved from one paddock to the next about every three days. The animals do not return to their original location for about a month or so. According to Sarah Gangahdin, a former pre-vet student of mine, if the eggs of *Haemunchus* egested in the feces of our sheep hatch, but the larvae don't find a new host within 14 days, they will die. So when my sheep return to the first paddock, their parasites are dead.

The other weapon in my arsenal is inspection. FAMACHA is a parasite inspection technique developed in South Africa that allows us to identify lambs (and adults) most likely infested with *Haemunchus*. FAMACHA is based on an examination of the lower gums and the conjunctiva, the membrane around the eye. In healthy animals, the lower gums and conjunctiva receive a rich supply of blood. They appear red on inspection. An infestation of the blood sucking *Haemunchus* reduces the amount of blood in circulation, causing the gums and conjunctiva to appear pale, even white, in severe cases. These animals, which in a healthy flock represent only a few percent of the total, need to be medicated. Thus, only a small proportion of the *Haemunchus contortus* population in the flock is treated and resistance to the medication rarely becomes a problem. In our flock, only two lambs were treated with anthelmintics this past year. The genetic diversity of my enemy, *Haemunchus contortus,* ensures its sensitivity to my attacks.

The farm represents the world in microcosm and the biodiversity associated with a farm is a reflection of the farmer's philosophy of production and interaction with the world around the farm. Traditionally, farms produced a wide variety of products — a seasonally varying diet for one's customers. Those products may have spanned a range of species and, within species, several varieties, breeds, or subspecies. Taken together, the products of a single farm reflected a respectable richness of species and, often, a vast genetic diversity. For instance, a single farm might produce several kinds of fruits, vegetables, pasture grasses (hay) and grains; as well as cattle, sheep or goats, horses, and various kinds of fowl (chickens, guinea fowl, turkeys, ducks, geese). Within each category there might be one, or a few, or dozens of species, varieties, or breeds. So a farmer might simultaneously be growing several kinds of apples, pears, and peaches; an array of vegetables and grains that varied seasonally; as well as cattle for beef and dairy; and sheep for meat, wool, and possibly milk. The farm was a diverse place and that diversity was capable of supporting the better part of a balanced human diet. A farm, in short, could sustain the entire person.

The vast majority of farms today produce one or two principal products. Ask a farmer what he or she produces, they'll tell you beef, or dairy, or specialty crops. The most common products of modern agriculture are feed grade corn and soy. There are people who spend their entire careers in agriculture producing only those two crops — thousands of acres of them. In the 1960s a farm in Schoharie County, New York produced only carrots. All of the carrots that this farm produced went to one place, the

Beech Nut Baby Food factory, in nearby Canajoharie. When Beech Nut decided to go elsewhere for carrots, the farm went broke. Today that same farm produces many kinds of vegetables, sold in both wholesale and retail markets. Richard Ball, the farm's owner, has a successful farm store that retails products from his farm as well as other farms and businesses. His inventory includes meat, milk, cheese, eggs, soap, candles, and cute bric-a-brac that people associate with the "country" — rural Americana. There is a greenhouse that Richard's late wife Sue started. At Christmas, Richard's daughters construct kissing balls, wreaths, and chains of pine boughs. Richard brings in locally grown Christmas trees, and the store is stocked with all the "country" ornaments you can imagine. One of Richard's daughters runs a sandwich counter and bake shop in the retail store. The bake shop gives the store a cinnamony smell that people associate with a country kitchen. I think that smell stimulates people to buy even more cute "country" bric-a-brac. Basically, the whole family is involved. And the place is profitable. While Richard still checks the price of commodity carrots coming out of Canada every morning (if the price is high enough he can sell a lot of carrots quickly in the commodity market), he is locked into neither a single product nor a single market. His is the business analog of biological diversity — he practices diversity in all facets of farming. And, it appears, the result of diversity — biological or economic — is stability, the ability to resist disruptions and disturbances, and to recover from their effects when, inevitably, they occur.

On the other hand, most farmers whose operations are dominated by a single commodity participate in a highly volatile market that seems to become less stable with time. These farmers sell their products to corporations, wholesalers, and middlemen. They work in a system of miniscule profit margins, enormous debt, and ridiculously low product diversity. If, instead of being farmers they were stockbrokers, the lack of portfolio diversity and stability would lead one to question their qualifications. As farmers, however, this is considered normal. The fact is, single-product farming is risky — a fast-track to failure. Diversity in any system creates stability while the lack of diversity leaves one exposed, with little chance of recovery from perturbation.

Historically, diversity in agriculture was promoted as the "natural order" of the farm. Land grazed by livestock would be enriched by their manure. That land would eventually be planted with crops that could be harvested. The gleanings would be ploughed back into the soil to feed the next year's crop, and so on. Eventually, the land would be rested, then returned to pasture for the grazers to fertilize and further rejuvenate. The landscape was sub-divided and each section played a different role. And the roles changed. One year the north fields would be producing crops; a few years later they would be feeding cattle, or producing hay. This was not a perfect system. To be sure, there were (and always will be) crop failures, overgrazing, and a whole range of disasters associated with miscalculations, missed opportunities, and the simple deterministic chaos that describes all systems close to nature. Ultimately, however, it was a system that worked, in part because of the diversity of production.

In modern industrial agriculture hundreds to thousands of acres might be planted in a single crop, which effectively "sets the table" for those insects and fungi that consume that particular crop. Billions of dollars are spent annually to chemically protect agricultural monocultures from the scourges that these so-called "modern" methods pre-ordain. And, of course, when we spray pesticides (biocides, really — there are very few poisons that kill only the pests and leave the "good" insects) they kill the pests...mostly. In almost every population there will be those few that are resistant to the pesticide. As these resistant members reproduce, with even more food available to them thanks to the eradication of their sensitive cousins, they chow-down to the banquet before them creating a new, pesticide-resistant, outbreak. The solution? More chemistry, of course. And so it goes, new chemicals constantly narrowing the gene pools of the pests. (Sir Albert Howard would call them "indicators of substandard practice.") In *The Ecology of Eden* (Penguin, 1996) Evan Eisenberg describes the long term outcome of this chemical approach to monoculture. In 1948, Eisenberg writes, farmers used 15 million pounds of insecticide and lost seven percent of their crop to insects. In 1990 they used 125 million pounds of insecticide and lost 15 percent of their crop. The unintended victims of this "wonders-through-chemistry" approach to farming are pollinators and other wildlife. Fields that used to be pollenated by wild bees and wasps that lived in the hedgerows and wood lots are now pollenated by rent-a-hive bees when available. More often than not, they are pollenated artificially. Everybody pays more to keep the industrial system running at "peak efficiency" (if you call that efficiency) while chemical manufacturers realize huge profits.

The campaign against agricultural biodiversity doesn't end with our post-war infatuation with chemistry. The industrial assault on farming has now been taken up by molecular genetics. Today we can clone single individuals that have attributes we desire, and we can implant genes that impart new attributes to our crops and livestock. We can implant genes that prevent pests from eating a crop (without having to resort to pesticides) and we can use crops as crucibles to produce medicines and other bioactive compounds. We can give plants the genetic capacity to resist weed killers, allowing us to spray a field and eliminate everything that is not wanted in it, and we can genetically modify the chemistry of a cow's rumen to reduce the production of methane, a potent greenhouse gas. Critics, however, point to new food allergies, invasions of domesticated species into wild ecosystems, and other pernicious outcomes of genetic modification. They suggest a liberal application of the precautionary principle, rather than unquestioned acceptance of these new transgenic technologies, absent a careful evaluation of their consequences.

A friend of mine, Peter Ten Eyck, grows apples at Indian Ladder Farms, near Albany, New York. He is what most people envision when they think of a farmer who uses sustainable practices. For instance, it's difficult to grow apples without using pesticides (even organic farmers use approved organic pesticides). Many apple growers apply a dose of pesticide as a matter of routine. But Peter applies pesticides only after pests are detected on his plants in sufficient numbers to do significant damage. Peter has a diverse operation, selling cider, pumpkins, and retail products, as well as many kinds of apples, in his farm store. His farm is available for parties, he has a petting zoo and an arts and crafts barn, and he has sold his development rights to a land trust, ensuring that his farm will forever be a farm. Peter Ten Eyck, however, is a vigorous proponent of genetic modification, "If you can add a gene to my apples that makes them repel pests so that I don't have to spray pesticides into the environment, I'm on board!"

So which is it — genetic modification (GM) or no GM? A significant part of the problem is that the question is being asked as if it had a simple "yes" or "no" answer. It doesn't. We, as a society, are not trying to understand the many grey areas of transgenics. A significant part of the problem may not be with genetic modification, but with those promoting the process.

Genetic modification, through selective breeding, is foundational in agriculture. Charles Darwin used the analogy of selective livestock breeding, which preceded his explanation of "natural" selection by many decades, to describe the mechanisms by which nature "favors" the most adaptive traits in organisms. Today we know that these traits are passed on by genes. Selective breeding, which maintains or increases the genetic diversity *within* species, conforms to some fundamental rules of nature. Only members of the same species can breed. Members of different species are reproductively isolated, meaning that the breeding members of two populations are prevented from mating by physical barriers such as mountains or oceans. Alternatively, members of different species may be able to mate but fertilization of the egg does not occur, or the fertilized embryo dies, or the offspring are sterile. A mule is the sterile offspring produced by the mating of a female horse with a male donkey.

Modern transgenic science changes that dynamic. Traits, i.e. genes, often from different species than the host, are inserted into the host genome with no mating taking place. Nature is not without examples of this mode of genetic modification. Viruses insert their DNA into the genome of their host, usually to the detriment of the host. Furthermore, transferred genes rarely do exactly what they did in the donor species. Nor is the desired trait the only one that is transferred with a gene — genes rarely code for single traits. Several "unintended" traits (as in unintended consequences) will almost always accompany the transfer. Furthermore, the process is expensive. To insert a gene from one species into another, one needs a well-equipped laboratory and technical expertise rarely available on a farm. That's why the process is usually performed in university and corporate laboratories which, incidentally, often vie for patents on the modified genomes they create.

For thousands of years farmers have performed genetic modifications by selective breeding. During the past century they have received help from agricultural colleges and Land Grant programs. But today farmers are out of the loop. Gene insertion is just too expensive and too technical. As a result, agriculture and, in fact, the American food supply (already dominated by a few multinational corporations) is now under nearly complete industrial control. A very few corporations own patents to the genomes of the seeds used to grow some of our most important crops. They have cornered the market on these crops. Farmers seeking non-GM seeds are often out of luck. Most Americans would be surprised at how thoroughly genetically modified products have been inserted into the food supply. In 2012, 90 percent of the soybeans and 80 percent of the corn grown in the US were produced from genetically modified plants. Farmer complaints about corporate bullying are nearly legendary. Academic researchers whose data contradict the industry position claim to have been harassed. Careers have been threatened when researchers have published findings that dispute corporate claims about the safety and quality of genetic modification. The federal government has taken positions that appear biased in favor of the corporations producing genetically modified products. In 2009, when Elena Kagan (now Associate Justice on the US Supreme Court) was solicitor general in the Obama administration, she submitted an apparently unsolicited amicus brief supporting Monsanto in a case before the Supreme Court, Monsanto v. Geertson Seed Farms et al., in which organic farmers in California were attempting to prevent the release of transgenic alfalfa.[2] Why she took the time to submit the brief, when there was no request to do so, is unclear. Associate Supreme Court Justice Clarence Thomas, a former counsel for Monsanto, refused to recuse himself in both Monsanto v. Geertson Seed Farms, and the patent infringement case of Monsanto v. Bowman. During arguments in the latter case, Chief Justice John Roberts wondered why people would do all the work of creating new seed strains if anyone can infringe on the patent and simply use the product. He seems to have forgotten that farmers have been developing new seed strains for thousands of years, and never patented a single genome. The very idea of patenting genomes seems rather ridiculous. The seeds or livestock that quantitatively or qualitatively produced the best yields made money for the farmer. The farmer didn't worry about excluding the public or other farmers from using the seed. Competing farmers simply strove to produce better-performing seeds. That's what has created the genetic richness and quality of our food supply. That's called *free enterprise.*

The patenting of genomes, the refusal of the industry to label GM ingredients in packaging, their reluctance to admit that safeguards to protect culturally pure products from contamination by GM pollen have failed badly, the bullying of farmers, and the bias apparent in a federal government saturated with former executives of the corporations that produce genetically modified products, all serve to create mistrust among consumers (and should). A big part of the problem with transgenics is not about the products as much as it is about the behavior of the corporations that control these products. Until these problems are resolved it will be difficult to have a mature and honest public debate about genetically modified foods.

If we ever do get to the point where a reasoned debate is possible, I would suggest a simple litmus test to judge the process: If a technology enhances species or genetic diversity, it will increase the stability of both agricultural and wild ecosystems. If the technology reduces diversity, it will destabilize these systems. This simple rule can help guide the agenda in agricultural genetic engineering. If we implant a gene that creates immunity to the effects of an herbicide in a single plant species, then plant the species in a field and spray with that herbicide so that all other plants are killed, we have greatly narrowed the diversity of the field. If we repeat this process over large areas, genetic engineering will add to the instability already created by industrial monoculture. Ultimately, herbicide resistant weeds will emerge (they already have).[3] Ultimately, GM plants will escape cultivation and become invasive in native communities (they already have).[4] Ultimately, genes intended to be constrained within one variety will contaminate others (they already have).[5]

The creative and carefully constrained use of genetic engineering has the potential to enhance biodiversity in agriculture and to provide nutritious food for our rapidly growing human population, much as farming has done for millennia. Nonetheless, transgenic product development cannot be allowed to proceed without transparency. The rigorous ecological and biological research required to understand the consequences of gene implantation must exceed the rigor of the market research intended to sell the product.

The key is motivation. Although fiduciary reward (money made) is always a consideration, when money becomes the principal driver of technological development, the quality and value of that development is compromised. I remember when the process of patenting genes first began. University professors, generally motivated by the quantity and quality of published scholarly research, and by the recognition they received for that work, discovered that they could produce significant wealth from the genes they patented. More than one researcher has suggested that the business of patenting genes has created a profit motive that has changed academic genetic research forever, not necessarily for the better. There is a clear parallel to agriculture here. Farming needs to be profitable. But when the triple bottom line is sacrificed at the altar of profitability, the operation fails to be sustainable. Profitability should emerge from the mission. It should not be the mission.

In the end, the farm is a transition zone between wild and urban environments, from soil to asphalt. The extent to which farmers recognize their place in that transition determines how the farm relates to the needs of the wild and human communities that converge at the farmscape. Farmers who reject that role tend to exclude both wildlife and the human public from the farms. These farms take on the look and feel of factories. During the 20th century many mammals, birds, pollinating insects and other arthropods, plants, and soil microbes have been lost to industrial agriculture.[6]

Increasingly, both farmers and consumers are questioning this approach, rejecting the artificial segregation of humans from the rest of nature, finding instead that we are inextricably embedded in nature and that the farmer is the gate keeper, occupying the middle ground between wildness and domesticity. As the industrial system collapses, as it must, the emergent agriculture recognizes the futility of trying to separate our species from the rest of the living world, our community from the larger natural community, of which — for better or worse — we are a part. The modern farmer shares space with nature, convinces nature to tolerate the agrarian landscape, and finds both greater security and profitability working with, rather than against the natural order. Management-intensive rotational grazing, mixed cropping and inter-grazing (grazing several livestock species together, or sequentially on the same ground), encouraging or at least not discouraging wild birds and mammals from nesting and denning in the farmscape, reducing or eliminating pesticides from the management protocol — these represent the willingness of the farmer to share the farm and, ultimately, to blur the boundaries between us and the rest of nature.

HUDSON SOLAR

Drill baby drill!

— Vice Presidential Candidate Sarah Palin, October 2008

7

Energy and the Future of Farming

"God put all the energy we need in the ground. All we have to do is dig for it," proclaimed a Member of Congress from Oklahoma during a recent Congressional hearing on American energy independence. He was, of course, referring to oil, natural gas, and coal — the fossil fuels. These deeply buried remains of organisms that populated the Earth's surface 100 million years ago contain enormous amounts of energy. For each BTU invested in mining petroleum, for instance, we get between 30 and 300 BTU of energy out.[1]

But really, is this what God had in mind? I can't say that I know what God is thinking most of the time but I am pretty sure that if the Congressman took a walk in an Oklahoma hayfield, or an Ozark forest, or one of that state's few remaining prairies, he might have a better idea about where God put all the energy we need. And it's a lot easier to get than digging for it.

Every blade of grass, every leaf on every weed and tree, every algal cell and stalk of moss is a solar collector. Plants use solar energy to convert carbon dioxide to carbohydrates (simple sugars) that provide their energy. The process, photosynthesis, turns sunlight into electricity, and electricity into chemical energy that, in turn, converts carbon dioxide to food — carbohydrates — for the plant.[2] Plants convert about a hundred billion tons of carbon dioxide to carbohydrates, annually.

We don't have to worry about running out of solar energy for about another billion years or so. Nor do we need to make deals with unfriendly governments to gain access to the sun's energy, or worry about allowing the sun to fall into the hands of terrorists, or concern ourselves with solar spills. Most of the houses in the United States could be producing 50–100 percent of the electricity they use with simple roof-mounted solar panels.

We can and should energize our modern food production system with solar and other renewable energy sources, as has been done for most of the past 10,000 years. To be sure, the sun will always energize agriculture. Through photosynthesis it fuels the growth of our crops and the forages that we feed our livestock. Historically, soil amendments consisted of composts and manures derived from plants, often passed through the guts of livestock and fermented in the barn for a season or two. The ultimate source of traditional inputs, therefore, is recent photosynthesis.

Although fossil fuels subsidized enormous growth in agriculture during the past century, the end of that era is fast approaching, and the growth and development it spawned is not sustainable with the fossil fuel supplies that remain. Many farmers realize this and, increasingly, they are returning to solar and other alternative energy sources in their operations. More farmers will follow these pioneers, as they lead agriculture into a more energy-efficient and "locally" scaled paradigm. Quite simply, the future of food production and distribution are incompatible with the long term realities of peak oil and the current instability of fossil fuel markets. Alternatives exist and they are already being adopted by farmers.

Substantial "solarization" of the United States could be accomplished in a just a few years by simply transferring the hundreds of billions of dollars spent on oil-industry subsidies to the federal and state agencies that already dole out funds to home owners and businesses interested in alternative energy. When Pam and I installed our first 5 kW solar panel system, New York State paid 45 percent of the cost of the project. When we put up an additional 4 kW, USDA gave us funds to help defray the costs. We didn't even have to ask for the grants; our installers did all the paperwork. This isn't pie in the sky. These programs are out there, they work, and they could easily be expanded by simply transferring federal subsidies from the most profitable companies on the planet to middle class Americans.

At the State University of New York at Albany, Dr. Richard Perez and his colleagues study energy. Dr. Perez understands the potential of solar energy as an alternative to fossil fuels. He has calculated the capacity of all existing energy sources to meet current and future global energy demand. His results are shown in the figures below.

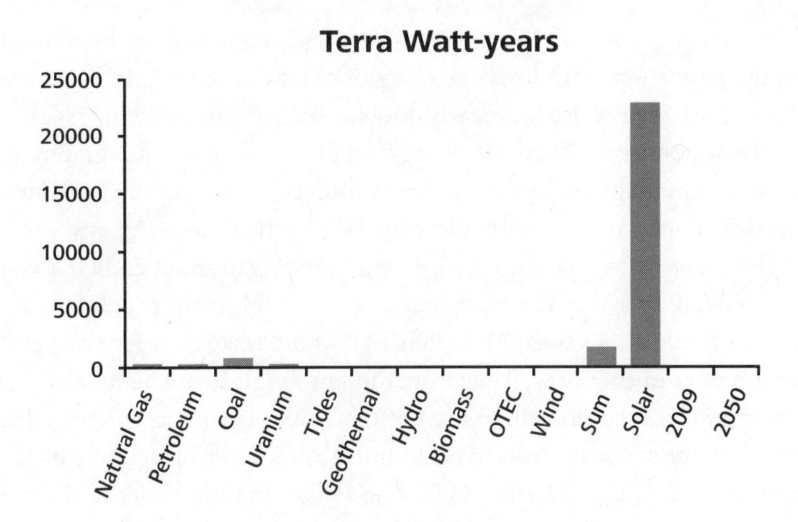

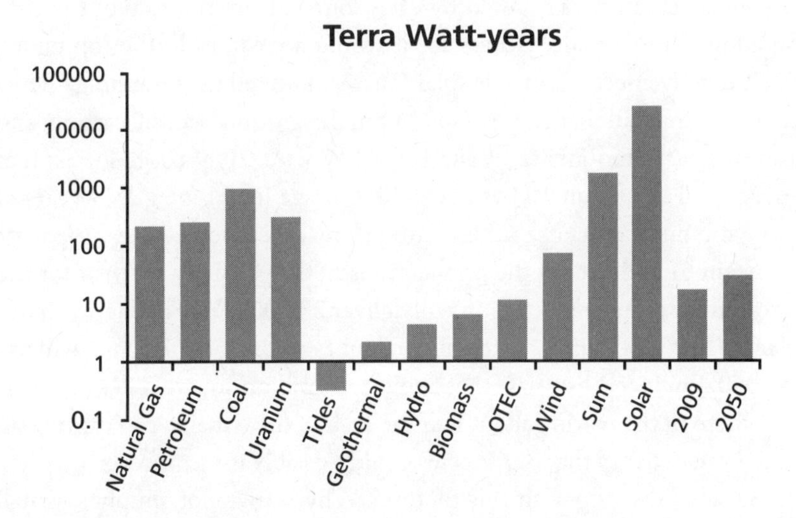

Global energy availability from non-renewable and renewable sources, as well as energy use in 2009 and projected use in 2050, shown in units of terawatt-years (TW-y). Top: *data plotted on an arithmetic scale.* Bottom: *data plotted on a logarithmic scale. Note: the bar labeled "Sum" represents all fossil and nuclear sources (755 TW-y). Drawn from the data of Perez et al. (2012).*

I plotted his data on two graphs. The units are terawatt-years (TW-y). A terawatt is a million megawatts. A megawatt is a million watts, or a thousand kilowatts. The electricity in our homes is metered in kilowatt-hours.

There are 8.76 billion kilowatt-hours in a single terawatt-year. In 2009, all of the people on earth used 16 TW-y of energy, according to Perez and colleagues. Energy use will nearly double, to 28 TW-y, by 2050.

How does that match the energy supply? Well in the top graph, the amount of energy available from each source (shown on the horizontal axis) is plotted on an arithmetic scale (the vertical axis). An arithmetic scale means that we count 1, 2, 3, 4, and so on, up to about 25,000 TW-y. Unless you look very carefully, it seems as though there is only one bar on the graph. That's solar. At 23,000 TW-y (and that is just for one year), the amount of energy available from the sun simply overwhelms all other energy sources combined. In the bottom graph, I've plotted Perez's data on a logarithmic scale. Instead of counting by ones, we're counting by factors of ten ... 1, 10, 100, 1000, 10000, and so on. In order to see how much energy is actually contributed by each source, I needed to plot the data on a logarithmic scale. Otherwise it would appear, as in the top figure, that the only energy source is solar. That should tell us something. Perez estimates that "all that energy God put in the ground" (fossil, carbon, and also nuclear) amounts to about 1655 TW-y. At 2009 usage levels, that energy will last about 103 years; at 2050 usage levels, we get 59 years of fossil and nuclear energy at best. Now, if we look at solar energy capacity, we begin to understand the implications of the graph. Every year for the next billion years or so, the sun will deliver 23,000 TW-y of energy to our planet. That's 821 times more energy than everyone on the planet will use in 2050. Engineers estimate that about two million square miles of desert anywhere in the world (about half the area of the Sahara) can capture all of the solar energy that our species could possibly use... forever.[3]

So what's wrong with this picture? Why do we not simply swap all that digging in the ground for some simple photovoltaics? Why do we continue to meet the major part of our nation's energy needs with oil, gas, and coal that are expensive, hard to get, dangerous to our health, and massively destructive to our land, water, air, and quality of life? Why are we not getting most our energy from the sun?

We hear lots of reasons why: We don't have a smart grid. We can't use the sun's energy at night or on cloudy days. We live in an apartment and the landlord won't put panels on the roof. It's too expensive — someday we'll have the technology to get cheap solar energy, but we're not there yet. But we all know (or should know) that none of these is the reason our nation has not committed seriously to the development of solar energy, even though we could.

The reason we produce most of our energy with fossil fuels, of course, is money. Our economy is built on fossil fuels. Too many people depend on ancient carbon for their livelihoods — extracting, refining, and selling fossil fuels, selling products made from and powered by fossil fuels, and investing in fossil fuel markets — to simply dump the fossil fuel economy and switch to an alternative energy economy. When politicians talk about transitioning to renewables, they're not talking about developing new fuel sources. Those already exist. They're talking about transitioning out of oil without collapsing the economy. How do you take a country bound to one energy paradigm and move it into another? Fossil-fuels form the backbone of economies globally. Those whose life's blood are oil, coal, or natural gas, will not go gently into a solar economy. There are people — powerful people — whose job it is to ensure that the world continues to depend for its energy on fuels that we all know are taking down our life support system. But they are not going to sacrifice their wealth and power for the alternatives. The bottom line is not about technology or even energy. It is about money — who makes it and how the distribution changes if we switch from fossil fuels to solar.

Agriculture is energy intensive. It requires that large amounts of energy be captured and harnessed in order to produce food surpluses. As a result, a relatively few people can feed many. The emergence of agriculture 8,000 – 10,000 years ago was transformative, changing humans from hunters and gatherers who in order to feed themselves were consigned to following the great herds of large herbivores, to settlers who could remain in one place and produce enough food for all of the people in their communities. Agriculture allowed the division of labor and that permitted the emergence of governmental structures, armies, cities, and states. In short, agriculture focused the sun's energy to fuel the creation of civilizations.

The introduction of fossil fuels to agriculture about a century ago brought an enormous increase in energy content to the system. Coal, oil, and natural gas ramped up the per capita productivity of agriculture and allowed the rescaling of the food distribution model. Along with its use as an energy source for growing and distributing agricultural products, fossil fuels were used in the production of fertilizer. The fossil fuel driven industrialization of farming initially improved the efficiency and effectiveness of agriculture, allowing a few individuals to manage thousands of acres, and yields to climb by orders of magnitude per decade. Led by biologist (and, later, Nobel Laureate) Norman Borlaug, the oil-energized "Green Revolution" sought to defeat hunger on Earth.

That grand vision, however, could never be realized. The escalating cost of fossil fuel extraction and purification (linked to their depletion and growing scarcity), government inability to subsidize those costs, the failure of synthetic fertilizers to maintain soil quality (as Albert Howard predicted), and the inability of living organisms to conform to an industrial model of food production have led agricultural economists to conclude that the age of industrial agriculture is over.[4] The overwhelming dependence of any food production system on fossil fuel is simply unsustainable.

Farmers find themselves caught between the market volatility of "peak oil" and resistance to any real commitment to a new energy paradigm by a handful of powerful companies and their friends in Congress. Yet many farmers who run small to mid-sized operations and never invested in the half-million dollar tractors and combines necessary to farm several thousand acres have been exploring, and often moving — independently and earnestly — toward alternative energy sources.

I doubt that it is possible to be commercially viable in agriculture today and not depend on fossil fuels to some extent. However, sustainability in agriculture demands energy security and that will not come from dependence on fossil fuels. It is abundantly clear that reducing dependence on supply-limited and economically destabilizing fossil fuels, and adopting alternative energy sources instead, is crucial to the future of farming. Scaling to direct and local retail markets, rather than to global commodities markets, brings the goal of energy security into focus. Like commodity food markets, commodity fuel markets destabilize agriculture. Scaling down from thousands of acres of production, to hundreds, tens, or even fewer acres makes it possible to reduce one's dependence on these destabilizing markets and to integrate alternative energy sources into the operation. As Wes Swancy, third generation owner of Riverview Farms in Ranger, Georgia points out, "If a man can't make a living off a couple of hundred acres of good land, the system's broke."

On my farm, we use a diesel-powered tractor. The most unpredictable part of my operation is powering that tractor. In the past year, the price of diesel (the most widely used petroleum-based fuel in agriculture) has fluctuated by nearly a dollar per gallon. It is difficult to set prices for what I produce when the price of a key fuel is as variable as diesel. My operation, like those of farmers globally, is susceptible to market fluctuations that threaten its stability and hence my profitability and even my sustainability.[5] Pam and I are actively investigating alternatives.

There is a retired engineer in Maine who has converted a small diesel tractor to solar. But the cost of that conversion is staggering and the power output of the unit is not impressive given its cost. One day a solar powered tractor may be under every farmer's shed, but even the engineer admits that it's probably a ways off.

A tractor is a device that pulls and pushes equipment to accomplish useful work. By that definition, there have been solar powered tractors on farms for thousands of years. They're called draft animals. During a conversation about using draft animals as an alternative to diesel, Chris Embler, co-owner of Applewood Acre Sheep Farm, exclaimed (clearly frustrated with my idealism), "What am I supposed to do, become Amish?" Chris runs all manner of diesel-driven equipment in his operation, including several tractors and bailers. "Well," I suggested, "Maybe the Amish can teach us a thing or two." The Amish scale their farming to their markets, most of which are local. Their approach reduces energy demand and allows the farmer more flexibility in determining the kind of energy needed to do a particular job. It allows them to use draft animals and human muscle for certain tasks that, at a larger scale, would require the higher energy content of diesel or other fossil fuel. Granted, it's not always easy to do with horse or human power what we have long done with our John Deeres, but by reducing the amount of land managed and by localizing markets it seems possible to begin weaning ourselves off the opiate of oil. This doesn't mean we need to go "cold turkey." By gradually reducing our dependence on oil, possibly by integrating a mule into the operation, or adapting bale-free haying techniques, the transition to the alternative energy economy can be accomplished at a comfortable pace (possibly with some benefits to our waist lines). Adjustments can be made along the way and we don't need to give up our beloved John Deeres completely. But it's nice to think that when we can no longer afford the cost of diesel — and that day will come — we can be sufficiently invested in alternative fuels to give up our petroleum-powered equipment and still be viable. The rate at which we abandon fossil fuels is equal to the rate at which we relieve ourselves of the burden of supporting an obsolete and dangerous industry. By focusing on local retail markets, a small to medium-sized farm — less than about five or six hundred acres — can feed several hundred families, using mostly draft and human power. I know this because Kristen and Mark Kimball are doing just that at Essex Farm. The Kimballs could probably double the number of shareholders in their CSA, and with it their gross sales, by increasing the use of fossil fuels in their operation. But profitability is not about gross sales. It's about the net remaining after accounting for the costs. Low overhead is what makes small-scale agriculture work. If the Kimballs are unable to feed everyone who would like farm fresh food in northeastern New York, perhaps we need more farmers. That's a good thing.

Can we feed everyone in the United States using half the fossil fuels that we do now? Not with agriculture "zoned" the way it is today. It is impossible to feed the majority of Americans without large trucks transporting food from one end of the continent to the other. I would suggest, however, that by refocusing the food distribution system to a more local scale, the majority of consumers could obtain significant portions of their diets at a much lower cost in energy. The fact that farmers' markets and CSAs are among the fastest growing sectors in agriculture is proof that local scaling works. Local scaling allows farmers to use alternative energy sources to power some of their operations. Not that the alternatives, such as draft animals or human power, are cheap — they're not. But the costs associated with maintaining draft animals or hiring farm hands are fairly predictable and less strongly driven by commodity markets. As the predictability of energy supply increases, the financial risks associated with farm operation tend to decrease.

Another alternative to diesel is the use of techniques with lower mechanical equipment requirements. That also reduces one's exposure to market fluctuations. This year, Pam and I are experimenting with high-grass pasture management. Normally, after our sheep rotate out of a paddock, we use the tractor to trim the vegetation. Different farmers will provide different explanations for why this is necessary. However, the Holistic School of pasture management holds that, other than the fact that the untrimmed pasture won't be as pleasing to the eye, there is no disadvantage to not trimming. In fact, high grass pasture tends to resist drought. This year, we are only using the tractor to trim electric fence lines in order to avoid short circuits.

Related to the matter of energy use on the farm is the issue of energy used to get products to market. Are thousands of farmers' markets, each populated by dozens of pickup trucks and hundreds of consumer vehicles, an improvement over long distance trucking? The question is complex. It is currently the subject of much research and the jury is still out on the answer. Rich Pirog and his colleagues at the Leopold Center for Sustainable Agriculture at Iowa State University have been studying the costs of local versus conventional food distribution systems for some time. Pirog et al. report that conventional trucking requires 4–17 times more fuel and emits 5–17 times more CO_2 into the atmosphere than the average local distribution system. At this point, however, there is no consensus about the impact of local versus long distance distribution models on energy efficiency.

Part of the solution to the problem of transportation may be the use of farm stands and farm-based retail stores. Because most consumers drive to a supermarket one or more times a week, the trip to a farm stand may not increase the total miles one drives significantly, if at all. Tom Daniels, a regional planner and author of *When City and Country Collide* (1999, Island Press), points out that more than half of the farms in the United States are located within 25 miles of a medium-sized city. Therefore, farms are reasonably accessible to consumers and, as an added benefit, people tend to enjoy the experience of visiting and shopping at a farm.

While alternative sources of energy, including wind and methane produced from manure, are increasingly displacing fossil-natural gas (which is principally methane), coal, and oil for power generation on the farm, photovoltaics are proving to be one of the most important players in today's agriculture. Above and beyond anything else, solar makes good business sense. Federal and state subsidy and grant programs are making solar panels increasingly affordable.

Richard Ball, at Schoharie Valley Farms, is a good farmer and a good businessman. He understands that if he produces his own energy he is going to save money. Last spring he installed a 100 kW solar electric system, which powers the refrigeration units that keep his produce cool after harvest. In all probability, Richard's son and daughters, who will someday inherit Schoharie Valley Farms, will never pay an electric bill.

Jim and Adele Hayes, at Sap Bush Hollow Farm, worry that the rising cost of electricity (about five percent per year) could negatively impact their successful business in pasture-raised meats. Jim and Adele, their daughter Shannon, and son-in-law Bob Hooper have worked hard to make Sap Bush Hollow a model of sustainable family farming. Their investment in a 21 kW photovoltaic system is helping them to stabilize the cost of running their freezers.

If this country is to have a future in agriculture — if another generation of farmers is to take up the yoke from the current, aging, generation — then there must be stability and profitability in the business of farming. This is impossible, or at least unlikely, when the future of agriculture is tied to fossil fuels. Realistically, few farmers are going to achieve complete severance from fossil fuels in the foreseeable future, but the less modern farmers depend on oil, gas, and coal the more stable, secure, and profitable their operations will be. Farming is and always has been about moving solar energy to the plates of consumers. The closer we come to accomplishing that photonic transformation without the help of fossil fuels, the more profitable farming will be and the more secure our food system will become. Farmers produce living organisms. The sun has been energizing life on earth for four billion years. Oil has been in the picture for about a hundred or so, with questionable results. Plants and animals can be produced well, in abundance, and profitably without fossil fuels.

In January 2012, I met with Kristin Kimball at the Northeast Organic Farming Association meeting in Saratoga Springs, New York to discuss the use of draft horses on the 600 acres that she works with her husband Mark and with the help of a dedicated and intrepid team of farm hands. She told me that they are still using tractors but that their goal is, ultimately, to have all of their equipment powered by draft horses and to have their tractors quietly rusting in the barn. I launched into my soliloquy about how we need to get out from under the yoke of big oil. I ranted about the economic instability associated with volatile diesel prices, and about how the rising cost of diesel is squeezing the profit out of agriculture. Kristin looked at me, perhaps a bit amused. "Besides," she said, almost in a whisper, "It really is the right thing to do."

APRIL AUDIETIS

If a farmer stopped farming every time he encountered a little bad weather or a little bad luck, pretty soon we'd all starve.[1]

— Carter Swancy, Riverview Farm, GA

8

The New Normal

On Sunday, August 28, 2011, Hurricane Irene crossed the warm waters off the coast of North Carolina and tore northward through the mid-Atlantic United States. By the time Irene reached New York's Hudson Valley she was nothing more than a tropical storm — but a very large, wet, and slow moving tropical storm. Irene thoroughly soaked the Hudson Valley and the Catskill Mountains, and then spread into the Schoharie Valley, western Massachusetts, and Vermont. The storm dropped massive amounts of rain, nearly eight inches in three hours, in the Schoharie Valley.

Although the sustained winds rarely exceeded 35 miles per hour, the first winds came out of the east. Because wind usually comes out of the west in eastern New York, the root systems of trees tend to be strongest on their west-facing sides. The east wind easily toppled thousands of trees barely supported by the rain soaked soil. Streams and creeks running through forests and fields, along roads and behind the old colonials and farmhouses that populate the region, quickly swelled and breached their banks, tearing away trees, brush, and roadbeds; and flooding homes and businesses. The saturated clay soils became impervious to water. The road in front of my home at Longfield Farm disappeared; not even emergency vehicles could navigate the river that only hours before was State Road 146 — an official evacuation route. As water accumulated behind dams the danger of a breach or worse, complete structural failure, necessitated the release of millions of gallons of water, exacerbating the emerging deluge in the Schoharie, and other creeks and tributaries of the Hudson and Mohawk Rivers.

The impact to communities that over hundreds of years had grown up along these waterways was predictably devastating. Within 24 hours, the agricultural region known to historians as the "Breadbasket of the American Revolution" was decimated. Fields of corn, three days from harvest, were flattened. As Frank Lacko and I walked through his corn field on the Schoharie Creek near Middleburg a few days later, Frank commented that even if he had had the time to harvest all of his corn and to get it into the silo, it would have been ruined anyway. He pointed to a silo — the corn inside soaked and fermenting — that had been picked up by the raging waters and moved about four feet off its stone base. "Can you harvest the corn as it lies?" I asked. "No way," Frank replied. "It's covered with silt from the creek. That stuff would destroy the teeth of my combine." In one sense Frank was lucky. He had recently moved his family to a house on high-ground which, along with his sheep barn which is also on high ground, fared reasonably well in the storm. Jim and Cindy Barber, owners of Barber's Farms, a regional icon just down the road, didn't do as well. Their entire crop and much of their infrastructure were lost.[2]

Jim Hayes, in Warnerville, had eighty sheep in a high meadow across Heathen Creek, about a quarter mile from his farm. The sheep found cover during the storm and were untouched by the flood below. Not everyone was so lucky, though. The worst of Irene was visited upon the Vanderplueg family that for four generations had selectively bred one of the best Holstein lines in the country. They locked the herd in the barn to ensure the safety of their animals during the storm. But they underestimated the ferocity of the flood. The barn, cattle and all, were washed away and have yet to be found.

People have been farming in the Schoharie Valley for three centuries. The topsoil along the creeks is deep and rich, replenished and nourished annually by spring freshets fed by the snow melt in the Helderberg Mountains. Farmers in the Helderbergs, which overlook the fertile valley, jokingly complain that the valley people are farming their soil. The rocky, yellow clay that remains is good for growing grass to feed cattle and sheep, but that's about it. The people in the valley are used to floods. Like their parents and grandparents they have profited from these periodic deluges, despite the loss of a house here and there, and an occasional road washout as the flood makes its way to the Hudson and Mohawk rivers, depositing topsoil and nutrients in its wake. Flood control is always on the agendas of state and local governments during the spring. Mostly, people know that floods come in the spring and, mostly, people have adapted to them.

Late summer and early fall floods have not historically occurred with the regularity of the spring freshet, and they are not as readily accepted by the locals. Late summer floods, like that caused by Irene, tend to "cleanse" the landscape at a time when it doesn't need cleansing. In mid to late summer the fields are full of corn and tomatoes, zucchini and snap peas. If the crop is lost farmers can't repay the loans they took in the spring to purchase their seeds. They may default. Those who are close to the edge of the financial cliff may be wiped out. That will be Irene's legacy.

Soon after the storm the state of New York and the US Department of Agriculture announced millions of dollars in low interest loans to help farmers get back on their feet. But farmers whose harvests have been lost and who won't be paying off their current loans are unlikely to take on additional debt. The failure of government to understand this limits their ability to help.

Farmers have played roulette with the weather for 10,000 years. They know the risks and they live with them. They tend to believe that if the weather beats them down this year, things will be better next year. For a farmer, who spends his days in the fields, there has always been a certain predictability to it all. After Irene, which was really the climax to several years of "odd" weather events — tornadoes, 80-degree days in March causing the apple trees to bloom, followed by killer frosts, intense hail storms, and drought — I detected something different in many of the old timers, who have been at it for a long time. I sensed a loss of confidence, as though the intuition gained from decades of farming had gone down the drain. There was a sense that the odds were changing, and that the game was less winnable. They seemed to recognize a new normal — that things might *not* be better next year. It is easy to become frustrated "harvesting" crop insurance year after year. "Bringing in one good crop of apples in four years," lamented Altamont Orchards owner, Jim Abbruzzese, "that's not farming." There is a growing recognition by those who spend much of their time outdoors that the weather we're used to is no longer the weather we're getting. The climate is changing and it's changing fast.

The problem is that climate change has become a politically charged issue, in part because remediation will require enormous changes in how we get and use energy. It will require significant lifestyle changes. The media and in many ways the scientific community has done a less-than-adequate job of explaining the strength of the scientific consensus on climate change. The dependence of the American political system on the good will of the industries most responsible for greenhouse gas emissions, as well as the failure of the media to explain the issue to the public, has led to climate-change denial. And this will prove deadly if our national mindset does not shift — and soon.

I have been a researcher for more than 30 years. I have studied the oceans, the coasts, and terrestrial landscapes — from planktonic food chains to suburban sprawl. I began reading the primary scientific literature seriously in 1974. In all this time I have never encountered a professional consensus on a major scientific issue to the extent that I have with climate change. There have been more than 13,500 peer reviewed scientific publications on climate change in the past decade, and only 24 cannot support the findings of the Intergovernmental Panel on Climate Change that (1) the climate is changing globally, (2) the rate of climate change is greater than would be expected as a result of natural forces alone, and (3) the rate of rise in average global temperature is correlated with upper atmospheric carbon concentrations, both of which have risen significantly since the beginning of the industrial revolution. That's a 99.8 percent consensus! Most of the 24 papers that do not support these hypotheses don't refute them either. They simply lack the statistical power to make a judgment, because in the short term, from one day to the next and from one place to the next, variability is enormous. This daily variability is called "weather." Only by collecting data about the weather over long periods of time — from pre-history to the present — and over a range of spatial scales, and treating that data with sophisticated statistical techniques, can we manage the variability and begin to see the underlying trends. This is not "meteorology for dummies."

Most of the people doing climate change research are math nerds. They are not affiliated with one political party or ideology over another. They are not trying to save the world, or conserve land, or do anything other than be the one who comes up with something new — a small improvement in the accuracy of a model, a new way to manage the variability, or an explanation for a certain phenomenon. That's a contribution.

Right now, every scientist in the world would give their eye teeth to be the guy who *disproves* climate change. That person has dibs on the next Nobel Prize in physics. Most of the scientists I know want as little to do with politics or the media as possible. Many don't believe that the media can ever get the story straight. And given the way climate change is reported in the media, that concern may be justified. Climate scientists, like most other scientists, deliver their findings to the readership of the peer reviewed, technical literature, not to the general public. Their work is intellectual, not social or political.

What the public and the media frequently fail to understand is that in science one never gets complete agreement. We get consensus at best, and weak consensus is the norm. Disagreement and argument are prerequisites. What is striking about the climate change debate is the degree of consensus. That doesn't mean that scientists agree on the details. How much? How soon? There are uncertainties — lots of them. But these are the details. Is it happening? Is it happening quickly? Is it human induced? The answers to these questions are settled.

Bottom line: When climate scientists tell us the climate is changing, and that we're making it change, we would be wise to listen.

That said, I predict we won't. There is too much money being made by sticking with the status quo. Too much of the current social fabric is wrapped around energy sources and behaviors that are based on the very things causing the global climate to change. While, for decades, nature has been sending clear signals that things are changing — ice sheets melting; 100- and 500-year floods every year or so; 30,000 Europeans dying in a heat wave — those governments that have been most responsible for the problem have consciously chosen to ignore the signs during much of the late 20th and early 21st centuries. Only recently did the government of the United States officially acknowledge the existence of climate change. However, the issue is such a political hot potato that it has been exiled to the back burner for much of the past decade.

It is not possible to decide if this or that meteorological catastrophe is due to climate change. The exercise is pointless, at any rate. There have been periods of stormy weather — years when the number of named tropical storms reaches "W" and years when we only get to "A" or "B." There have been drought years as well. The problem is that the "gain" is being turned up — the intensity and, in some cases, the frequency are increasing dramatically. New York State is not particularly prone to tornadoes. Between 1950 and 1995 an average of six tornadoes touched down in the state each year. However, from 1990–1995 the annual average jumped to 15.[3] Tornado activity has remained high in New York, at 10 per year, through the first decade of the 21st century. As I write this, 90 percent of the United States is in drought — 60 percent in extreme drought. In 2012, the American heartland experienced the worst drought since the 1930s — a considerable portion of the national corn crop was lost. In an average year the United States experiences three or four significant weather events (i.e. catastrophes that result in more than a billion dollars in damage). In 2008 we set a global record: eight significant events. In 2011 that record was shattered as we experienced 15 — more than any other year in recorded history. In and of itself, a severe drought or a couple of out-of-place tornadoes are not necessarily bellwethers, but taken together the events of the past few years are sending a signal that is hard to ignore.

The bigger problem is that large, complex systems like the earth-atmosphere system do not respond immediately, or necessarily proportionately, to changes in forcing factors. There are time-lags of various lengths to any change or perturbation. Such lags are common — we've all encountered them. MIT professor John D. Sterman makes the point by using the analogy of a bathtub filling with water. The tub represents the atmosphere; the water, greenhouse gases. When the water is turned off and the plug pulled, the bathtub does not empty instantly. It takes time. If the water continues to run, even slowly, it takes longer for the tub to drain. That is how CO_2 in the atmosphere behaves.

It is critical to appreciate the effect of time-lags if we are to understand how greenhouse gases will affect the earth's climate in the future. In the year 2000 the atmospheric CO_2 level was approximately 360 ppm, considerably higher than the natural background level of 228 ppm. If, in 2000, we had completely stopped using fossil fuels (all emissions had ceased — totally unrealistic, of course) atmospheric CO_2 levels would be back at 350 ppm[4] in or about the year 2050. On the other hand, if we continue adding greenhouse gases to the atmosphere at year-2000 rates, atmospheric levels will reach 440 ppm by 2050 and more than 500 ppm by 2100. At those CO_2 concentrations, the climate may be incapable of supporting human life. Unfortunately, the use of fossil fuels increased during the decade following 2000, accelerating the rate of CO_2 accumulation in the upper atmosphere. If current models are accurate (and their predictions have been pretty good so far) our species is in for a wild ride.

Clearly, any progress toward the recovery of historical CO_2 background levels resulting from our remediation efforts will take years. Certain climatic "tipping points" have already been exceeded — some elements of climate-past are no longer recoverable. Experts tell us that, as the global average temperature continues to rise, we are heading full bore toward several other tipping points, which once exceeded could be fatal. Will our civilization come crashing down? Will our species survive the floods, the droughts, the fires, the crop failures, the storms, the expansion of deserts? Nobody has the data to answer these questions, but civilizations great and small do crash — ours, if it crashes, will not be the first. In a recent paper, Professors B.L. Turner II of Arizona State University's School of Sustainability, and Deborah Lawrence, of the University of Virginia, showed how the physical structure of Mayan civilization — the ways they modified the environment with their cities and their agriculture, and the way they distributed their burgeoning population — was the result of engineering and management decisions about the environment that locked them into the very outcomes that ultimately brought down their civilization.[5]

A similar premise was advanced by Jared Diamond in his book *Collapse*.[6] Diamond suggested that decisions made about the environment have historically determined the success or failure of human societies. The reach of these decisions, however, have been reasonably local — on the scale of city-states or geographic regions. The consequences of climate change during the current century will be global in scale. Significant human-induced environmental changes have never before occurred at this scale. Our species will be required to adapt globally. That has not happened before. In past centuries, if a civilization crashed in Europe, there were other human societies elsewhere, unaffected by the changes their European "cousins" had brought upon themselves. The adaptations that will be required to deal with the current changes in climate will be different in different parts of the planet but the scale of requisite adaptations will be global and it will affect the entire human population. The adaptation of agriculture — our food production systems — must take priority. There is no other choice. The failure of agriculture guarantees the failure of human societies and, potentially, of our species.

But *how* do we adapt? How does agriculture adapt? How will farmers deal with hurricanes, mega-storms, tornadoes, droughts, and invasions of new pests that will emerge as warm isotherms migrate rapidly poleward? There is no simple answer. Suffice it to say that farmers are used to environmental instability. Farmers are accustomed to dealing with unpredictability, though they don't always do a good job managing risk. But risk management will be key to any viable response to future climate instability. I believe that three general principles of risk management will emerge: (1) Production (i.e. what is produced, where, and how) must be diversified, (2) Debt must be minimized and managed, and (3) Connections between farmers and theirs communities must be strong.

Diversification. Even in the best of times, monoculture is unsustainable. Climatic variability and uncertainty, radical swings between flood and drought, and heat and cold, make monoculture unrealistic for farming in the future. However, diversification does not simply mean that the farmer will grow a variety of crops. Rather, farmers will diversify the locations of their activities as well. Farmers will no longer plant their entire crop in the flood plain, despite the quality of the soil there. A portion of the crop, say a third, will be planted on high ground. These "safe" plantations won't provide flood-plain yields, they may require more inputs and be tougher to harvest. But in the event of a late-season flood, they will be more likely to survive. They will allow the farmer to provide food to customers and to see some financial return on investment, even if the margin is reduced.

Farmers who grow several kinds of products reduce the risk of loss due to failure of any single product, much as a balanced financial portfolio reduces the risk of losses due to volatility in the financial markets. While requiring expertise in several areas, rather than a single product, diversification makes sense in a variable and unpredictable climatic (and economic) environment. If one's row crops are lost to a flood, cattle grazing on the high ground will put food on the table and keep the farm above water. Multi-cropping and rotational agriculture, in which livestock and vegetables are rotated through the same landscapes, help distribute both costs and risks. None of this is rocket science. But it works.

Diversification will also occur in marketing, particularly among mid-sized farms. While large farms may be the last to leave the commodities markets, small and mid-sized operations will find it difficult to survive in a single market in the future. Coupling a bit of wholesale with multiple direct marketing strategies cushions a bumpy economic road and provides a safety net against market volatility. For example, Clemens Mackay and Jenny Rosinski of Solstice Hill Farm in Cobleskill, New York, were making ends meet with their CSA and their involvement in several farmers' markets. In the summer of 2012, they planted a field of tomatoes for a new farmers' market spinning up in nearby Cobleskill. Unfortunately, that market collapsed before it opened leaving Mackay and Rosinski scrambling for a buyer for their tomatoes. They barely made it through. The following spring, Jenny captured a small wholesale contract for vegetables, the two young farmers rented some additional land, and together with their existing direct marketing operations Jenny and Clemens are digging themselves out of the hole they were in.

Debt reduction. Wendel Berry wrote that debt is the farmer's worst enemy. Debt exacerbates the already increased risk to the farm brought about by climate change. Debt-risk is acquired by the need to take loans for buildings, heavy equipment, seeds, and inputs. Many farmers take a loan each spring to cover the cost of seeds, fertilizers, and pesticides. If all goes well, the loan is paid off after the harvest. But if the crop is lost to a flood or drought, the funds to pay off the loan may never materialize. The possibility of bankruptcy can become very real, very quickly. Selling the farm or a part of it may be the only option. Reduction of financial exposure by minimizing or intensively managing indebtedness is a prerequisite to farm viability in the face of increased climatic variability.

Community supported agriculture, CSA, is uniquely suited to debt management in an unpredictable world. The CSA model spreads the risk of catastrophe across a community of shareholders — stakeholders, really — who invest in the farm by "pre-buying" shares of the harvest. So instead of filing for a loan in the spring, Cathy Newcomb, the farmer to whose CSA we belong, sends us an e-mail in February letting us know the price of a share of the harvest from the Newcomb's Farm CSA and the mandatory pay-date. Along with each of Cathy's other shareholders we send a check and Cathy's husband Slim uses those funds to buy the seeds he'll need to grow our food for the season. If all goes well, we'll be enjoying an abundance of fresh vegetables from June through October. If disaster strikes, whatever losses are incurred by the farm are spread among all of the shareholders in Cathy and Slim's CSA. We share in the harvest and the hardship. Because it is so important, I have devoted an entire chapter of this book to Community Supported Agriculture (Chapter 12). Suffice it to say here that the CSA model, in its many and evolving forms, is likely to become a foundational element in agriculture in the future. It will take on new dimensions. It will be expanded to include all manner of farming techniques and products, and scaled according to a variety of markets. It will become a critical risk management tool, possibly *the* risk management tool in 21st century farming.

Community connectivity. A few days after Hurricane Irene had passed, I visited Richard Ball, whose farm sits on the banks of the then-flooded Schoharie Creek. The parking lot of his farm store was packed. When I found him in the store, Richard looked shell shocked. His entire crop had been wiped out, though he said he would be okay. He had set up a clothes distribution center in the flower nursery. The store was full of food, some grown on his farm and harvested before the storm, and some purchased from other area farms that had survived Irene's wrath. There were plenty of fruits and vegetables, handmade soaps, milk from a local dairy, grass-fed beef and pasture-raised pork, books on how to plant a garden and how to bake bread, spices, lithographs of pictures painted by a local artist, good coffee and cake, and so much more. "My community is gone," Richard said, in a broken whisper. "Then who are all these people?" I asked. "They're not from the Village of Schoharie." "Well then Richard, it looks like your community is bigger than Schoharie." If there was anything like a positive outcome from Irene, it was that it helped us all to appreciate the relationship between farms and the communities of people they support. The reach of a single farm into the community is often underestimated, until, that is, an event like Hurricane Irene occurs.

In farming, one's community (as shown in the figure below) consists of the farm, neighboring farms, businesses and institutions that support the region's agriculture, and the market — composed mostly of customers within the farm's "reach" who do regular business with the farm. As the paradigm of farming continues to shift, the emerging local approach to food production and distribution, which directly connects the farmer and consumer, will continue to develop. As it does, the bond holding the farm community together will grow stronger. Connectivity develops through farmers' markets, CSAs, food hubs, restaurants, and other establishments and institutions that source their food locally. The connections forged in the community and in these direct markets create a self-sustaining interdependence among the elements of the community. These relationships are more than financial. They are based on the respect that each member of

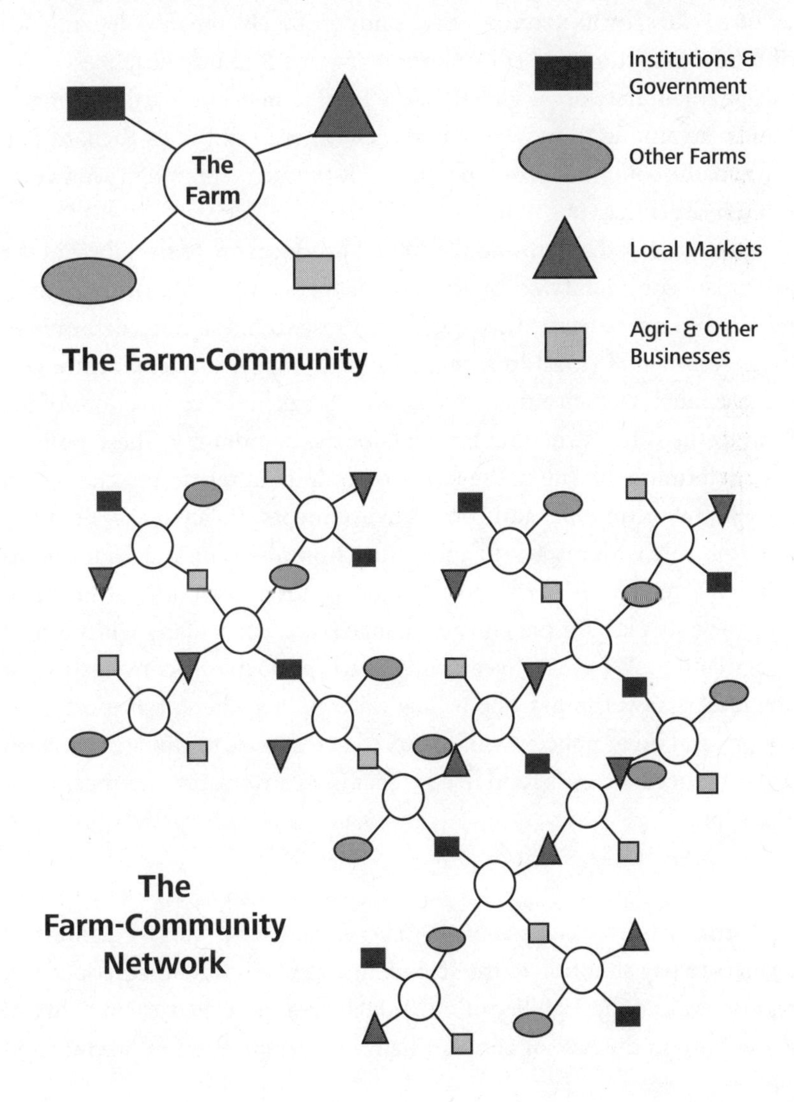

the community has for the other and on the central role that the farm plays in the community. Ultimately, as farm communities are interconnected, lattices or networks are created, much as the bonds between the atoms of a salt create a crystal (see figure). The evolving model of agriculture represents part of an emerging socioeconomic system that provides food security in the face of uncertainty. The community provides the structure that creates this security and the promise of sustainability both for the farm and those who depend on it. The community of the emergent agriculture is local by its nature — the bonds created by the interdependence of food production and consumption are weakened by distance — but the emergent network is global. In the local farm economy, the strongest bonds are among those who know each other best. But as Richard Ball learned, the community, or more precisely the farm-community network, is often larger than one perceives it to be.

As central as the farm and the food it produces are to the fabric of the proximate community, so are the strands that connect the farm-community to the larger society. The strength of those strands is often determined by governmental (local to federal) support of agriculture. Government creates policies which either strengthen or weaken the connections of the farm to the other structural elements of the community. These policies may determine the fate of the farm-community in relation to changes in the natural, economic, and social environments. Policy-makers tend to see their role with respect to agriculture from the economic side and to a lesser extent from a public health perspective. Government needs to recognize its place in the climate change issue, particularly with respect to agriculture. We are not yet having nearly enough of a conversation, at any level of government, about how water management, transportation, energy, and other policies will be altered in response to climate change in order to meet society's food needs. That is a conversation we must have, and soon.

Farming is a risky business — always has been. The willingness of farmers to pay attention to the science and to act in ways that reduce risk will determine the viability of individual farms and the sustainability of agriculture in the face of climate change. Diversification of product and process, reduction of indebtedness, and community-based management of land and water will prove critical to the ability of farmers, farms, and farm-communities to adapt to environmental change. At the end of the day, independent of climate, and regardless of where "normal" is, these are the principles of the emergent agriculture in the new environmental reality. The extent to which they are embraced will determine the future of farming, and indeed, the future of our species.

Part III

The Local Economy

PAUL TICK

Innovations that are guided by smallholder farmers, adapted to local circumstances, and sustainable for the economy and environment will be necessary to ensure food security in the future.

— Bill Gates, CEO, Microsoft Corp.

The re-establishment of farmers' markets in the United States gives me hope.

— Raymond Saul, just an ordinary guy

9

The Emergent Market

On September 20, 2011, just after noon, I was sitting in a third-floor conference room in the Russell Senate Office Building in Washington, DC, as a member of New York State's Agriculture Working Group. Richard Ball was explaining how row crop farmers make a living. "Every spring," he began, "You take a loan and you buy your seeds and your fertilizer. You plant and tend and harvest your crop. You sell your harvest, pay off your loan, pay your help and your mortgage, and maybe put a little in the bank. That's how agriculture works." He continued, "When a natural disaster takes out your crop three days before the harvest, you have nothing left. You can't pay your loan. You can't pay your help, or your mortgage, or anything. You're finished." This was about three weeks after Hurricane Irene ravaged the agricultural landscapes of the Hudson, Mohawk, and Schoharie Valleys of New York State.

The members of the Agriculture Working Group had been listening to USDA representatives describe the various loan programs available to farmers to help them pick up the pieces. To my left sat a large, ruddy-faced fellow — an onion farmer from the "black dirt" region of the lower Hudson River Valley — with a stoic expression that concealed deep pain. "How the hell am I supposed to take another loan?" He whispered. "I already owe two hundred thousand. What... I'm supposed to ask my wife to go to the bank and borrow another hundred grand? Can't do it." I realized that he probably would not be back for next year's working group meeting.[1]

Enter Thomas Christenfeld. With his wife Liz, Thomas owns The Alleged Farm in Valley Falls, New York, in rural Washington County. Thomas grows excellent vegetables that he sells through a CSA —

Community Supported Agriculture — program. Every April, each of the 350 or so members or member-families of The Alleged Farm's CSA writes a check to Thomas for about 500 dollars. Thomas uses that money to buy his seeds, pay his help and the mortgage, and put a little in the bank (if the tractor doesn't break down). Every week, from June through late October or early November, each "shareholder" gets a box of vegetables. Usually, the boxes start out pretty light — some greens and lettuces. By late June, however, the boxes are burgeoning with all manner of colorful, textural produce. And along with the produce comes Thomas' weekly newsletter. Thomas was trained as a writer (with degrees from Harvard and Stanford) so the newsletters are well crafted and usually full of humor (Thomas' writing style reminds me of Mark Twain). Embedded within the prose are subtle messages about the importance of family and farming, about how the unpredictability of the weather drives farmers crazy, and about the inevitability of weeds. There is always a description of the vegetables in the week's share and some recipes (usually involving olive oil, garlic, and a skillet). All-in-all, shareholders come away with well over 500 dollars' worth of vegetables, and some pretty inspiring reading.

A few years ago Thomas hurt his back. He had real trouble working and ultimately he needed surgery. The boxes of vegetables were light that season but, in the end, Thomas paid his bills. The following season the boxes were once again full. The Alleged Farm had come through the crisis — not without pain, but with farm and family intact. Thomas' shareholders had shared more than vegetables that year. They had come to appreciate what the "community supported" part of CSA means. That is part of the reason why the Alleged Farm made it through. Thomas' injury might have been devastating had The Alleged Farm functioned on the conventional model. As a CSA, however, the impact of the disaster was spread among the shareholders and, in the end, the Christenfelds and all of the families for whom they grow food still had a farm from which that food would continue to come. The CSA is a hallmark of the new approach to farming that recognizes, first and foremost, the importance of managing, and where possible, eliminating debt. Secondarily, the paradigm shift in agriculture is characterized by a focus on direct, retail markets. Wholesale and commodities are certainly components of many operations, particularly mid-sized and large operations, but direct marketing is a cornerstone of the emergent system. The CSA model addresses both elements — debt reduction and direct marketing.

Thomas Christenfeld's outcome is far from typical. Often the farm does not recover from a medical crisis or an environmental disaster. If income depends on an unpredictable commodity system or a vertically stratified payment system controlled by large, distant corporations, the potential for bankruptcy increases dramatically. As the spatial or economic distance between producer and consumer increases, the viability of the operation becomes less certain and the control of outcomes by the farmer decreases.

The survival of small to medium-sized family farms depends on the ability of farm-families to escape, or at least to conduct an organized retreat, from markets that are fundamentally unstable. This was apparent to the earliest practitioners of the now burgeoning community of farmers who practice some form of direct marketing. It is grounded in the premise that factors which increase the risk of failure must be abated. As obvious and logical as this premise seems, it has been difficult to achieve and has sometimes been ignored, with predictable consequences. Risk management is a strategic pillar of the emergent market.

Profitability is key to sustainability. Without profitability, farming is either a hobby or a path to bankruptcy. When the price paid for a product is less than the cost of its production, the producer is on the road to bankruptcy. As ludicrous as this sounds, under-payment is part of the fabric of commodity-based agriculture (see Chapter 3, Farm Subsidies). For instance, in 2005, the cost of producing a bushel of corn in Iowa was approximately $2.50. The price paid to the farmer at the grain elevator was $1.45. Federal subsidies based on the commodity price of corn are "theoretically" used to make up the difference. However, corn subsidies, and with them the farmer's profit margin, have declined in recent years. The problem is that if the farmer doesn't like the price being offered at the grain elevator, he can't go to another elevator. Prices are pretty much fixed by the few multinational corporations that own the elevators. This is called a bottleneck. A similar bottleneck exists in the beef industry. Four corporations process 81 percent of the beef in the United States. They own the feedlots and the five or so large packing plants that process most of our beef. They buy the rancher's stock prior to delivery of the cattle to the feedlot. Between 1982 and 2002, the price of beef paid to ranchers fell from $0.63 a pound to $0.46 a pound. That price often did not cover the cost of production. Not coincidentally, the suicide rate among cattle ranchers is three time the national average.[2]

I have a friend who produces milk. I met him a few years ago, after delivering a lecture on the paradox of agriculture. He told me that he was 30 years old, has a wife and three kids, and an Ivy League education. He milks 500 Holsteins twice a day and is $7 million in debt. "The bank won't even let me sell my land. They just keep giving me money." "How much are you getting for your milk?" I asked. "Fourteen dollars per hundred-weight,"[3] he responded. That was the commodity price. "How much does it cost you to produce?" "Sixteen dollars per hundred-weight." It didn't require an Ivy League education to see that my friend is going broke. At about this same time, a gallon of milk in the supermarket was going for about $3.[4] The industry has determined that a standard gallon of whole milk weighs 8.6 pounds. Using this weight, at the supermarket a hundred-weight of whole milk would go for $34.90, two and a half times what my friend was getting, and about 218 percent of what it costs him to produce. Stating it another way, my friend was getting about $1.20 per gallon, 40 percent of the supermarket price. Why is there such a sharp difference between the price at the barn and that at the retail market? Well, there are a lot of hands on that milk. There is the company that trucks the milk to the pasteurization facility, and the wholesaler who buys the milk and sells it to the retailer — who then sells it to the public. In the end, just three companies — Dean Foods, Kraft, and Leprino Foods — control about 70 percent of that process.[5] They don't own the farms. Instead, working together with large cooperatives, they determine what the farmer gets for their milk and what the public pays on the other end. The National Family Farmers Coalition has reported on the manipulation of milk pricing by collaboration between large dairy cooperatives and distributers such as Dean Foods.[6]

Why doesn't my friend simply sell his milk directly to the public? He can't. The law says that to sell milk off the farm it must be pasteurized. To do that would require a large investment in equipment, facilities, and labor — far more than a man $7 million in debt can afford. My friend is caught in the pasteurization bottleneck that helps a few companies to maintain their control over milk in America. Congress has tried to help farmers caught in this bottleneck by increasing the subsidies for dairy farmers being paid less for their milk than the cost of production. But is that really a help? Or is it just life-support — a way to temporarily sustain an unsustainable system?

The commodity food chain is too long. There is the farmer, the wholesaler, the retailer, and ultimately — calling the shots — is a board of directors. My friend is a superb craftsman. I could not get 500 Holsteins milked twice a day if my life depended on it. But in the end, nobody knows what he does. Nobody appreciates the craftsmanship involved in being a dairy-man. He is part of a black box, comprised at one end by our iconic images of farming — of black and white cows grazing happily on a hillside — and at the other end by a clean crisp carton or plastic jug of milk in our grocer's cooler. Nobody knows his name or the name of his farm or even how much of the milk in that carton is his. Really, nobody cares. His milk meets certain government standards for bacterial count, butterfat content, and so forth, just like the milk from a hundred other dairies that is picked up and transported to the pasteurization plant and then distributed to supermarkets, anywhere.

I bake bread for a farmers' market. I can produce about 130 loaves of bread in about 24 hours. I sell those loaves for about $6 each. I generally sell all of my bread in less than three hours. It costs me about two dollars for the ingredients in most of the breads that I bake. Other than the costs of running my ovens and transportation, I'm clearing $4 (200 percent) per loaf. What are my customers paying for? Well... they're paying for me. It's not just my time. They are paying for my skill as a baker — as an artisan. The ability to make a sourdough starter, mix the pre-ferments, manage the three ovens simultaneously, and oversee the development of as many as twenty doughs at a time is a craft. My customers understand this. They appreciate not only the time I spend baking for them, but also the time I spend explaining the process of making bread to them at the market. Many know my name. After my farmers' marketing season ends, several come to the farm on weekends, some driving more than 30 miles, to buy a couple of six dollar breads. Some of my customers have become our friends. One, a professional editor, actually edited this book for me... gratis! People get it! They want farmers to succeed, for farms to be profitable.

To state the obvious, profitability is achieved only if the farmer produces a product for less than the market is willing to pay for it. Profitability improves by reducing the number of steps between farmer and consumer, and the cost of each step. That means direct marketing, particularly for small-scale and many mid-sized operations. Direct marketing through farmers' markets and CSAs is the fastest growing sector in agriculture according to the USDA. Creating a market at the farm is even more efficient, if the location is right. Farm stands and retail stores on site obviate transportation costs. And for the mid-sized farm, too small to compete in commodities and too large to sell only at farmers' markets, diversification — from wholesale, to retail, to catering — is critical to profitability. Jim Abbruzzese loves producing apples at his beautiful Altamont Orchards. But he would have gone broke long ago if he and his brother Joe hadn't put a golf course right through the orchard, and built a restaurant, and set up a small farm store and a nursery and, recently, a catering business. The Abbruzzeses are a commentary on both the business savvy of the American farmer and their love for the land.

The downside of direct marketing is that the producer has to take the time to sell his or her products. Some farmers have no problem with this. But others prefer being farmers to sales reps. Many just don't have the time. Farming is a "team sport" and family members, interns, and field hands — just about everyone but the dog — have to participate in marketing the farm's products. My experience at farmers' markets is that the salesperson matters. Whoever is doing the selling must know the product and the process used to produce it, because market patrons care about these things. And, of course, the seller must enjoy selling. The energy the seller brings to the market correlates directly with sales. The tent at the farmers' market is a retail environment; its appearance counts and direct marketing requires commitment to... well, marketing.

The market can be challenging and the proliferation of farmers' markets around the nation requires that farmers make some fairly sophisticated decisions about what, where, and how to sell. On what day of the week and at what time is the market? How many other vendors are selling the same product? Do the vendors get along? What is the cost of space?

Are there unusual insurance requirements? Is it a seasonal or year round market? Can I do more than one market per week? Who monitors quality? Are crafts people and fast food vendors involved? And, of course, how much traffic is there at the market location and how much parking is available? For years I participated in a market in a beautiful small village a few miles from my farm. The market was readily visible from the road and had excellent access for foot traffic, but signage announcing the market was scarce, as was parking. Vendors did not do particularly well.

Every week, Jody and Luisa Somers, owners of Dancing Ewe Farm in Granville on the New York-Vermont border, transport their sheep and cow's milk cheeses (created from ancient Tuscan recipes) to New York City's iconic Green Market at Union Square in lower Manhattan. Along the way, they drop off cheese at some of the city's best restaurants. On Saturday, Dancing Ewe is at the markets in Saratoga Springs and Troy, and on Sunday they are in Rhinebeck in the Hudson Valley — a lot of driving, but the return justifies the cost. As you pass the Dancing Ewe tent in Saratoga Springs, Luisa gives you a glance and whispers in her wonderful accent, "Would you like to try some cheese?" A sliver of caciotta passes into your fingers, then to your mouth. And you're hooked. You will not walk away from the Dancing Ewe tent without at least a quarter pound of that incredible cheese. Everything about Dancing Ewe speaks to the marriage of superb craftsmanship with skillful marketing. Jody and Luisa understand that you need both. A great product without a way to reach the market will get the family a lot of good food and the farm a path to bankruptcy. A great marketing strategy without the quality to back it up will get you exactly one day of good sales. Jody and Luisa understand that sustainable farming practices must sustain both the land and the farm family.

Coincident with the growth of farmers' markets has been an explosive increase in the popularity of community supported agriculture. In the past 30 years the number of CSAs in the United States has increased from two to more than 8,600. More than any other marketing approach, CSA reduces the financial risk of farming. From its roots in Rudolf Steiner's anthroposophy school of philosophy in early 20th century Europe, the community concept emerged in the 1960s as the marketing strategy for the organic and biodynamic farming movements — as an alternative to the already suspect industrial food system. The first CSAs opened in

North America in the 1980s; they have spread prolifically since. CSAs, such as the one operated by Jean-Paul Courtens and Jodi Bolluyt at Roxbury Farm in Kinderhook, NY for the past 30 years, have built large and loyal followings. Roxbury Farm has more than 1,000 shareholders. And farmers like Jean-Paul have trained dozens of young farmers in organic technique and in the use of the CSA model to increase financial stability. Today, 60 percent of Wisconsin's family farms that sell through the CSA system obtain their entire family income from the farm. That is a striking indication of success. A new breed of farmers is experimenting with and pushing the limits of the CSA model, testing new approaches to distribution, increasing the variety and quantities of food available to shareholders, and stretching the season to as much as 12 months. I discuss the exciting and constantly evolving CSA model later in the book.

While marketing is ultimately the farmer's job, the yoke is not solely the farmer's to shoulder. State and federal agencies have been impressed by the successes of direct marketing in agriculture and are now committing resources to programs that foster small-farm viability. And the market as well must be willing and accessible — the consumer must be interested in the product and the farmer must be able to reach out to the consumer at a cost that permits profitability. Consumers must recognize their responsibility to reward craftsmanship and hard work with patronage, and be willing to pay a fair price for their food. The examples above suggest that they are. That's what makes this time so interesting. All of the pieces are in place. Consumers are actively seeking out farmers' markets, CSAs, and other opportunities — to buy directly from the producer, to know the life cycle of their food, and to be assured that the food they are buying is safe. Supermarkets have taken to listing the names, and even displaying pictures, of the farmers that produce their vegetables. High-end catalog companies, such as Williams Sonoma, will sell you lamb, cheese, and even bread from farmers and artisan producers whose names and pictures appear in their catalogues. An emergence is underway in agriculture, not just in the way farmers produce food, but in the way people think about their food.

The books, films, media reports, and undercover videos documenting the inhumane treatment of livestock, the filth, the dysfunction, and the general insecurity of the American food supply are all beginning to have an effect. The assumptions that all food is the same, that the only thing that matters is price, that we can avert our eyes as animals are treated in ways they should never be treated, that we can trust the industry and government to ensure food safety, no longer hold water. The public is waking up. People are disgusted with what they're seeing and they're looking for alternatives.

Of course, the industry isn't oblivious to what's going on and there's a noticeable increase in the corporate cultivation of "fuzzy feelings" about farming, with heartwarming pictures of fathers and sons and family dogs on the backs of orange juice containers, and pictures in the coupon section of the Sunday paper of farm families gathered 'round the table... all very touching. But as they say in Kentucky — "that dog don't hunt!" You can't patent the genomes of the American food supply and call yourself sustainable. You can't put a hundred and twenty-five thousand chickens in a house and call yourself humane. You can't pay a farmer less for his milk or his grain or his cattle than it costs him to produce it and call yourself ethical.

Americans are fed up with the food system. They are fed up with being part of artificial commodities markets — with a system where eating a hamburger may be suicidal. People are seeking new markets and they are finding them right in their communities. These are markets where, for a reasonable price, consumers can buy food that tastes the way it is supposed to taste. These are markets where the person who hands you a loaf of bread actually baked it, where the person who hands you a bag of carrots has dirty knees because he picked those carrots at five o'clock that morning. These are markets where the person who hands you a package of ground beef is going to feed that same beef to her family that very night.

While I don't believe in fate, I do believe in inevitability. I also believe that all living things instinctively follow paths that they perceive will lead to greater viability. When one discovers that the path one is on, a path thought to be sustainable, is in fact separating them from their life support system, it is inevitable that they will seek a new path. That is what is happening in the American food system. The emerging market is a new path. It offers hope for the future of farming. It is inevitable that, shown the path, we will take it.

Pam Kleppel

The farmer is the only man in our economy who buys everything at retail, sells everything at wholesale, and pays the freight both ways.

— John F. Kennedy

10

The Consumer in a Changing Food System

I BAKE ABOUT FORTY DIFFERENT KINDS OF BREAD. These include a variety of sourdoughs and yeast loaves, rolls and baguettes. I make whole wheat bread with flour ground on our farm, from a heritage wheat strain grown only in New York State and Canada. I bake pain au levain and chef Chad Robertson's "Tartine" style breads in a wood-fired beehive oven, an olive loaf that one customer said is better than the olive bread his mother used to make when he was growing up on the Isle of Cypress, a white bread made with raw milk from my neighbor's dairy, garden-herb infused focaccia, semolina filoné, wood-fired bagels, Irish brown bread, soda bread scones, and Welsh cakes. I sell exclusively at farmers' markets and at our farm. I have been at it since 2004. I almost never return home with bread. I sell out — usually 130–140 pieces in about two hours. That's better than a bread a minute! I realize this measure of success, in part, because I'm a pretty good baker and I market the bread well. People crave good food, made with high quality ingredients. I provide that. I also provide information about how the bread is produced. People want that information. But there is something else that determines my success — something that emerges when I run out of bread. Rarely do those who arrive after I've sold out grouse about how I need to bring more bread to the market next time. Rather, the responses to my spent inventory are congratulatory: "Good for you!" or "I'll have to get here earlier next week." These comments affirm the public's commitment to a system they consider important — local production works when consumers are committed to the system. That's a powerful message. People get it. They realize that they are not simply buying bread, but that they are purchasing a piece of the local food system — *their* local food system.

True, these people are shopping at a farmers' market. They are the choir. We are singing from the same hymnal. According to some they are the gentry — elitists. But I sell bread for food stamps too. Given the success of farmers' markets over the past decade, we may be limiting our perspective if we claim that participation in these markets is restricted to a certain class of consumers. From where I stand interest in small-scale, local food production is not limited to particular social classes. The only thing that's limiting is the amount of bread I can bake or the number of tomatoes the guy in the tent next to mine can pick before loading his truck for the trip to market. People rich and poor, pushing strollers, and being pushed in wheel chairs rise a little earlier on Saturday mornings so that they can get to the market to get some really good bread, really nice tomatoes, and the many other products that the market's artisans have to offer. They place value on the quality of the food they eat and on the system that produces it.

Profitability is determined in the market. Consumers must value products sufficiently to pay a price that exceeds the cost of production and transportation. Price is determined by myriad factors, which include the larger economy, the scarcity of the product, and the importance the consumer attaches to it. Farmers who work in the emerging market produce food for a clientele that believes good food is worth the time required to seek it out. If necessary, most are willing to pay a bit more to get a better product. This modern consumer is increasingly aware of the life cycle of their food and is becoming educated in the language of the production process.

Although the intent of those producing food in this emerging system is simplicity and proximity to the earth — whole foods, minimal processing, minimal use of external inputs and alterations — the market is complex. The consumer receives an enormous amount of information about food, much of it from an industry whose interest may be best served by making information about food production complicated, deflecting the debate away from quality and safety, and refocusing it on price and convenience. The consumer is left to make an enormous number of decisions about products and production pathways with which they may have little or no experience. How far, for example, should I as a consumer travel to find the kind of food that I want? Conversely, how far should my food travel to get to my plate? Is certified organic spinach grown on an enormous farm in California, picked and washed mechanically, packaged in plastic and transported 3000 miles in a refrigerated truck, better than spinach grown two miles from my home by a farmer who uses synthetic fertilizers, and pesticides when necessary? We are deluged with opinions about how to think about these questions from advocacy groups, chefs, and academics who write books. Ultimately, however, each of us will find our own answers. The information needed to answer these questions — and, more importantly, the ability to verify the information one gets — is readily available within the emergent system. That is where the new agriculture and the conventional system diverge.

While nominally intended to provide information and confidence in agricultural products, the lexicon of food has become a source of confusion and, sometimes, misdirection. The terminology is staggering. Labeling often clouds rather than clarifies consumer understanding of the production process. Labels create (not necessarily by accident) undeserved illusions about how a particular food is produced. For instance, the term "free-range" has become associated with a method of producing poultry that is both humane and, somehow, nutritionally superior to conventionally-produced poultry. However, USDA "free-range" poultry may be neither of these. The USDA definition of free-range for broiler chickens ensures neither humane treatment of animals nor nutritional quality.

The most common "broiler" chicken (i.e. birds raised specifically for meat) is the fast-growing Cornish-cross. It requires only eight weeks from egg to freezer. Other, less popular chickens, such as the Freedom Ranger and myriad heritage breeds, grow more slowly and produce less meat, particularly breast meat, which is what consumers often want these days. According to USDA standards, a broiler can be labeled as "free-range" if it has access to an "outdoor setting" for at least two weeks (about 25 percent) of its life.[1] Tens of thousands, even hundreds of thousands of free-range birds may be packed into houses, usually under artificial light, for 75 percent of their lives (about six weeks). When the doors finally open, the birds are reluctant to leave. They can be "free-range" if they want to be, but they generally won't want to be when they've been inside for three-quarters of their lives. We began to understand free-range from the point of view of the chickens as a result of an incident on our farm a few years ago.

We raise broiler chickens. The chicks arrive at the post office (yes, they are sent by regular mail) a day or so after hatching. For the first three weeks of life the chicks need to be kept in a warm place. We keep our chicks in the barn, under a heat lamp. At three weeks of age, while they are fledging (getting their adult plumage) they are moved to our chicken coop, where the space per chicken is sufficient to ensure that birds, when fully grown, will not be crowded.[2] Each morning we open the door to the coop and I move the chickens' food outside to encourage the birds to join me. They do. And once there, the chickens begin scratching and pecking at the ground, supplementing their grain-based diet (chickens are omnivores — grain is a natural part of their diet) with bugs and grass and flowers.

A few years ago we received our first shipment of chicks in the first week of April. The last two weeks of the month were unusually cold and rainy. We were forced to keep the chicks in the barn for five, instead of three weeks. During that time, stress in the flock increased and one of the birds was killed in a fight. Another died of a heart attack. When we were finally able to open the door to the coop not a single bird would move. They wanted nothing to do with the outside world. Pam and I had to lift each chicken out of the coop each morning. Even with this effort, we were not satisfied that the process created what we would consider a truly free-range bird (and we explained that to our customers). Nonetheless, by the USDA standard for free-range chicken, we were compliant. And while our valiant effort was feasible with a couple of dozen birds, it's not even worth discussing in an industrial operation with a couple of hundred thousand. Our experience makes the point that meeting the USDA definition of free-range does not guarantee the consumer a bird that has ever been on pasture and may in fact deliver a product quite different from what the consumer thinks it is.

Another term, "cage-free," refers to chickens that lay eggs. Industrial "layers" are routinely confined in cages about the size of battery boxes. In fact, they are called "battery-box" chickens. Studies have shown that "battery-cage" confinement prevents the bird from performing normal bodily movements, such as turning around or spreading its wings. It shouldn't require a study to recognize that this process is obscenely inhumane. A cage-free chicken is simply one that is not confined in this manner.

"Cage-free" is not a UDSA designation. It does not imply that the bird has been provided with its biologically-required separation (the pecking distance) from the next bird. It does not imply that its diet consists of anything more than a manufactured powdered grain-mash or that other needs that most of us would consider basic animal care are being met. It simply means that the chicken is not living in a cage the size of a battery box.

A few years back the staff at *Mother Jones News* sent eggs from 14 farms on which the chickens were allowed to roam freely or were moved around the pastures, to a certified food testing laboratory for nutritional analysis. The staff at the magazine compared the analyses of these eggs to USDA egg nutritional data. The average *true* free-range egg contained one-third of the cholesterol and one-quarter of the saturated fat of the average USDA egg. The free-range egg contained 67% more vitamin A, three times more vitamin E, two times higher omega-3 fatty acid levels, and seven times more beta-carotene than the USDA egg. The difference is clear — but not at a supermarket's refrigerated egg case. There, the consumer gets the message that the cage-free eggs came from chickens that were not confined. Sometimes there are cardboard posters next to the refrigerated case, picturing chickens wandering contentedly on hillside pastures. Yet one cannot verify, at the supermarket, that the birds in the picture produced the eggs in the refrigerated display case, or that they lived a decent, uncrowded life, or ever saw the light of day, or felt grass and earth beneath their feet, or could behave like chickens rather than egg production units in a factory.

If the label doesn't reveal the true production process what does? The answer is simple. The answer is 'the producer.' Verification of the process can only be provided by the producer. The consumer must be able to speak with the producer and, if one so desires, to see the process first hand. At the supermarket the best that one can do is to read the label and believe what one will. In consumer-direct markets, such as farmers' markets, CSAs, or the farm itself, one can usually talk directly to the producer — one should be able to look at the birds and the facilities used to maintain them. On our farm customers can watch as I move the "chicken tractor" — the hen-house-on-wheels, in which our egg-laying chickens sleep and lay their eggs. Every week I move the chicken tractor to a paddock that our sheep have recently vacated. The chickens follow the sheep. That mimics nature. As the wildebeest and zebras migrate, birds follow the herds, scavenging for parasites in the dung, scratching the manure into the soil and revitalizing the land. When our sheep leave a paddock, in their rotation cycle, the chickens move in. The birds eat the parasites left in the sheep dung, and scratch the dung into the ground, helping to fertilize the soil. By simply allowing our chickens to *be chickens* the quality of our pastures improves. The chickens seem quite content with all of this, as evidenced by the fact that they lay eggs all year long, even though we don't artificially increase the photoperiod during the winter. (It is widely held that

G. Kleppel

chickens require a photoperiod of about 14-hours for egg production. It is unusual for the birds to be producing eggs in mid-winter without artificial lighting, but they do.)

The labeling of beef, lamb, and pork can be even more confusing to the consumer. Many consumers are familiar with the terms grass-fed, pasture-raised, and naturally-raised. What does all of that mean? What is the difference between them? Can pork be grass-fed? Is a pasture-raised cow the same as a naturally raised cow?

The term grass-fed means exactly that. Except when necessary to protect the animal's health the diet is restricted to grasses and forbs (weeds) foraged from the pasture during the growing season, and to hay when the pasture can no longer be grazed (in temperate climates that is generally late fall to early spring). Hormones are never administered to sustainably-raised livestock, whether grass-fed or not. Antibiotics are administered only when an animal is sick. The cost of high quality hay (more expensive than grain) and the extra time required for grass-fed livestock to complete the growth cycle, increase the price of the product.

The final stage of livestock production is called "finishing." During the finishing process the animal generates a layer of fat around the meat that imparts flavor and texture. At best, American cattle and sheep finish slowly on grass. Many will not finish on grass at all. The energetic content of grains, particularly corn (often supplemented by hormones and other growth-enhancing substances during industrial production) is considerably higher than that of grass. The combination of grain and chemicals cause rapid growth and fat deposition (much of the fat, unfortunately, consists of artery-clogging, low-density lipids), allowing the animal to finish quickly. The widespread use of grain feeds and chemical supplements has also resulted in genetic erosion in livestock. That is, the genetic capacity to finish efficiently on grass or hay has been lost in most breeds (see Chapter 5). Many farmers are working to recover the genes that will allow their livestock to use grass efficiently for growth, lactation, and other key processes. This too, is discussed in Chapter 5. Suffice it to say here that, eventually, the ability to finish rapidly on grass will be re-established in American livestock. This will allow the cost of finishing on grass and hence, the price of grass-fed meat, to more closely match the costs and prices of meat from the industrial system.[3]

While the term grass-fed refers only to cattle, goats, and sheep, the term pastured or pasture-raised can be extended to swine and fowl. Pigs, turkeys, chickens, ducks, and geese are omnivores. They eat more than grass and hay. As such, pastured pork or fowl production implies that the livestock are not confined, or force fed grain, i.e. that they are essentially free ranging animals with access to pasture and to a variety of foods. As with cattle, goats, and sheep, hormones are never used to alter growth rates or stimulate weight gain in pastured swine or fowl, and antibiotics are administered only to sick animals. There is, however, no government standard for pastured meat.

The American consumer has been growing increasingly skeptical of the quality and safety of the food sold through conventional markets — and with good reason. The trade journal *Food Production Daily* reported that every year approximately 76 million illnesses in the United States are linked to tainted food.[4] In 2005, the US Center for Disease Control and Prevention in Atlanta reported that more than 325,000 hospitalizations and 5,000 deaths were caused by contaminated food. In 2004, 2005, and 2006, an average of two out of every thousand samples of ground beef in the US contained unacceptably high levels of *E. coli* O157:H7, the strain of the *E. coli* bacterium that causes tens of thousands of cases of pathology and 61 deaths in the US each year.[5] It has been reported that the FDA has allowed contaminated foods into the market, knowing that they were tainted.[6] Imported food has the worst record in the American market and the amount of food imported into the United States is increasing.

The growing interest in local, organic, and humanely produced food is, therefore, not surprising. The impressive growth of consumer-direct markets in the US, where these products can be purchased and where one can talk to the producer, speaks to the concern of consumers about food security. According to the USDA the number of farmers' markets in the United States grew by just over 18%, from 3,706 to 4,385, between 2004 and 2006, and by 250% in the decade between 1996 and 2006. While this kind of growth cannot continue indefinitely, the top of the trend is not yet in sight.

Concerns surrounding the conventional food system are not confined to contamination and health issues. It should be obvious at this point that a significant part of what consumers pay for in the market has nothing to do with the food itself. Rather, we pay the cost of getting that food to market. This is not a single cost, but rather a system of costs that involve transportation companies and a series of "middlemen" who purchase food at various points along the "chain" and re-sell it, ultimately, to retailers who then sell it to the public. Each step adds to the cost of the food — while never adding to its quality.

In 2007, the Transportation Business Association (TBA) estimated that the cost of the diesel fuel to deliver food to the nation's plates comes in at hefty $139 billion annually. That's about $250 million a day. That figure, however, is based on a diesel cost-at-the-pump of $2 per gallon. As I write this the price of a gallon of diesel is above $4, potentially doubling the cost of getting food to one's plate without any improvement in its quality. Markets began responding to increased transportation costs with 6 to 20 percent increases in the price of produce, eggs, and juice between June 2006 and June 2007, according to the US Department of Commerce.

Direct markets, farmers' markets, CSAs, and farm stands provide alternative means of getting food to consumers, and a different cost structure. Twenty years ago, relatively few consumers were willing to pay the extra cost associated with the production of safe, nutritious food. Fewer still could afford to regularly purchase artisanal food products. The so-called "niche markets" were populated by chefs from the very best restaurants, food service professionals who appreciated the value available from this alternative food system, and a stalwart corps of advocates, who questioned the quality and sometimes the safety of food from conventional sources.

A new demographic is emerging, however. It is composed of a portion of those consumers who, until recently, believed that products bearing labels such as "locally-produced" and "organic" were fads and not worth their actual or perceived additional costs. Such skepticism is rational if food is really a commodity and if food can be assumed to be safe. This is exactly what the industry wants us to believe. However, the public is coming to the realization that food is not a commodity and that food safety cannot be assumed. These factors, coupled with the increasing cost of food from the industrial system, are making the alternative market more appealing to middle class consumers who, until recently, were not convinced that the so-called sustainable diet was worth it. It turns out that with minor modifications to household budgets food produced in a responsible manner is quite affordable. If current trends continue, about two-thirds of the American public will soon have access to locally, ethically, and safely-produced food.

If local food production and marketing continue to grow in popularity, the survival and even growth of small-scale farming is virtually assured. This won't necessarily come easily. Multinational agribusinesses will not go quietly into the night. As increasing transportation and processing costs drive food prices up, protests over food prices of the sort seen in the 1970s may erupt again, and corporate lobbyists will push government to step in with "big solutions" to "help" consumers — and corporate bottom lines. In the past these "big solutions" have led to greater food insecurity, an explosive increase in small-farm bankruptcies, and (of course) enhanced profitability for large corporations. However, since the farmer in the industrial system is working about as cheaply as he or she can, the American food system, arguably, has nothing more to give up in order to lower food prices. Fortunately for the industry, treaties such as the North and Central American Free Trade Acts (respectively NAFTA and CAFTA) along with similar agreements with governments in Asia and elsewhere, will likely allow the cost of production to be pushed even lower to offset rising transportation costs. What will be compromised, unfortunately, are food quality and safety. The offshore shift in food production will ratchet down overall food security and this, I predict, will be the undoing of the industrial food system.

In the 1970s, the public accepted government assurances that the food provided to American citizens was safe. Today's consumer does not accept that premise unconditionally. However difficult it is to ensure domestic food safety, it is orders-of-magnitude more difficult to assure the safety of imported food. Consumers already know this. At some point, the safety of one's family becomes more important than a few cents off the price of a pound of ground beef. That will be the tipping point from which the globalized industrial food system will not recover. An array of forces, including increased personal wealth and an explosive growth in the number of locations where sustainably produced food can be conveniently purchased, are combining to bring middle-class consumers into the emergent markets.

To control the market, multi-nationals have created bottlenecks in the movement of products from the farm to the point of sale. As I described in Chapter 9, a bottleneck is created by legally limiting the path that a product can take as it moves to market. Whoever controls the bottleneck, controls the market. For example, dairies cannot sell milk directly to consumers unless they pasteurize their milk on site. Pasteurization equipment is expensive. Therefore, most dairies sell at unrealistically low commodity prices to co-ops that pasteurize the product and sell it to a wholesaler or retailer. The rules governing the milk market are made by Congress, and large milk co-operatives and corporations spend enormous sums of money lobbying members of the House and Senate in order to retain control over that market.[7] In 2005, more than $5 million was spent on campaign contributions and lobbyists to close a "loophole" in the western milk market that permitted a single, independent dairy operation to sell milk directly to the retail market at a lower price than the cooperatives.[8]

Small-scale producers, who have been bottlenecked by agribusiness and commodities markets, have used a variety of creative or just plain "gutsy" approaches to generate profitability. Dairy farming, like other bottlenecked sectors of agriculture, has been hard hit by negative pricing impacts. Some dairy farmers, who can afford the equipment or can get a loan to buy it, have begun pasteurizing their own milk, allowing direct marketing and obviating control by the co-operatives. Alternatively, some dairies are producing raw milk cheeses. Others have replaced "high milk volume" Holsteins with lower volume Guernsey, Swiss Brown, and Jersey cattle that produce milk with the high butterfat content desirable in ice cream. Niche-appeal is increased by becoming a certified organic producer. The contaminant-free milk, cheese, yogurt, and ice cream from organic dairies are actively sought by pregnant women, families with small children, and people who prefer dairy products free of the toxic contaminants present in conventional milk today.

Just as sustainable agricultural practices have emerged as the alternative to industrial farming, so have alternative marketing approaches allowed these new farmers to get their products to consumers. Some 19,000 farmers obtain their entire agricultural income from farmers' markets and tens of thousands more participate on a part-time basis. Farmers' markets take many forms, from a few venders selling produce from the backs of their trucks on a few weekends during the summer, to juried year-round affairs that operate in permanent structures. Older markets, such as the Reading Terminal Market in Philadelphia, PA, which opened in the late 1890s in a building (called the shed) that was specially constructed by the Reading Railroad at its Market Street Terminal, have become consignment-based supermarkets.[9] The Pike Place Market in Seattle offers space to farmers and fishers, crafters, small-scale retailers, and street performers.[10] Some 28 markets are operated in New York City by the Green Markets Program, which has been developing farmers' markets in the five boroughs since 2002.[11] More than 200 producers participate in these year-round markets. Different markets are open every day of the week.

The determination of what actually constitutes a farmers' market is a bit fuzzy and is often determined by the vendors or by the sponsoring group. In general, farmers' markets should allow direct sales of locally produced farm products to the public. Many farmers' markets sell both basic produce — vegetables, fruits, and meats — as well as some value added products, such as bread, cheese, preserves, and vinegars. Farmers' markets often permit crafters to participate, particularly those whose products have agricultural themes. For instance, wool and flax spinners, weavers and knitters, and painters and photographers whose work deals with agricultural and rural subjects are often welcome at farmers' markets.

Perhaps the most important and rapidly evolving model in the emerging direct-to-consumer market is Community Supported Agriculture, or CSA. The CSA approach has helped thousands of small to mid-sized farms become or remain viable by making it possible for farmers to reduce the risk associated with growing food — by using shareholder funds provided before the growing season to purchase seed and livestock, instead of taking a loan to cover these purchases. Equally important is that CSAs create bonds between farmers and consumers that contribute to the formation of strong, stable farm-communities and food networks.

In the end, viability in modern agriculture is about the relationship between the farmer and the consumer. It is a relationship based on ethical production and honest communication by the producer about the processes used to produce food. Perhaps more important, however, is the consumers' enlightened appreciation of the effort involved in, and the value created by, such practices. Each of us, as consumers, should realize the critical role we play in agriculture. The emergent system celebrates the consumer as a partner. The bonds created at the interface between producer and consumer — the farm and the market — are central to the sustainability of the emergent food system.

When I lecture about agriculture I usually start by asking, "How many of you live or grew up on a farm?" A hand or two might go up in the audience. "How many of you are currently involved in agriculture?" Again, there may be a hand or two in the air. Finally, I ask, "How many of you have eaten today?" Of course, everyone raises a hand. "So if you've eaten today, you're involved in agriculture." As Wendell Berry put it, "Eating is an agricultural act." "What you eat," I tell my audience, "sends a message that is heard by farmers around the world!" Agriculture could not exist without consumers. The consumer sets the agenda for agriculture. The farmer implements it.

G. KLEPPEL

We must learn to invest as if food, farms, and fertility mattered.

— Woody Tasch

11

Slow Money

In 1985, the McDonald's Corporation opened its first store in Italy — in the town of Balzano. Almost overnight, golden arches erupted across the Italian landscape. Not surprisingly, this attempt at fast-food-ification in one of the most gastronomically sophisticated cultures on earth was quickly and loudly rebuked by the emergence of the Slow Food movement. As a counterpoint to the industrial fast food juggernaut, Slow Food celebrates local cuisine. It insists that the ingredients used to prepare food be safe, nutritious, fresh, and flavorful; that eating be a social process; that meals should be enjoyed with others. Most importantly, the Slow Food movement demands that food production be ethical. Within a few years, the movement had re-affirmed the longstanding food cultures of Italy, France, and Spain and stimulated greater appreciation of well sourced and prepared food in the United States.

The philosophical foundation of Slow Food is that the earth will not be poisoned, the air and water will not be polluted, and livestock will not be abused by our food system. Slow Food rejects the model of a meal as something consumed without thought. It speaks to family values, to ethics, to spirituality and to the health of the individual, the community, and the nation. With its focus on economic, cultural, and ecological values, Slow Food sends a message that is irresistible — that food matters. The principles of the Slow Food movement have been embraced by restaurateurs, farmers, conservationists, animal rights activists, clergy, and family values organizations.

Emergent from the Slow Food movement, in 2009, was the concept of Slow Money, introduced by venture capitalist Woody Trasch in his book *Inquiries into the Nature of Slow Money: Investing as if Food, Farms, and Fertility Mattered.*[1] Slow Money is a way of thinking about the qualitative outcome of investments. In finance, we are usually most concerned with the quantitative outcome — we invest money with the expectation of getting more money back. What Trasch argues for is investment in community and, in particular, one's local food community. The return on such investments is measured in increments of quality of life rather than quantity of accumulated capital. Slow Money argues for the diversion of capital from global markets to local markets — to farms, farmers' markets, and restaurants within one's own foodshed. Slow Money creates an ethic around one's personal investment strategy. Trasch argues that nurture capital must replace venture capital, that carrying capacity and care of the commons must replace greed, and that satisfaction with the quality of one's life must replace abject materialism. Like Slow Food, Slow Money celebrates craftspeople — farmers, bakers, cheesemakers, and chefs — who focus on making a living by offering quality goods and services, often with both passion and conviction. This has never been a recipe for making a killing. As the financial crash of 2007-2008 illustrates, in recent times money has been too fast, corporations too big, and investing too complicated to be sustainable in the long term. A more personally attuned relationship with capital provides outcomes not obtainable by high-stakes venture capitalism.

Not surprisingly, Slow Money has been criticized as wide-eyed idealism. But why must everything that does not exalt greed be wide-eyed idealism? Slow money won't change Wall Street any more than Slow Food will do in the Big Mac. But the Slow movements do represent alternatives to a financial industry that provides enormous benefits to the few and insecurity to most, and a food system that is grossly destructive to our health, to our concept of social equity, and to our planet's life support system.[2]

At the heart of the Slow movements is the simple but essential truth that food is not a commodity. A commodity is an industrial product. Wherever a commodity is produced, it is produced in pretty much the same way. As such, its value fluctuates as a function of externalities — that is, from factors not associated with how it is produced. For instance, a bar of gold used for trading and investment is called a kilobar. It weighs

PAM KLEPPEL

exactly 1,000 grams. Gold bars are produced in pretty much the same way everywhere. There is nothing in particular that distinguishes a gold bar in Fort Knox from one in a vault in Marseilles. The value of a bar of gold is determined by factors, such as political instability, that are not associated with the production of the gold itself.

The question is whether food has these same commodity attributes. When produced in an industrial environment, on so-called factory farms, food increasingly becomes a standardized good — a tomato is a tomato; ground beef is ground beef. By physically separating the producer and the production process from the consumer, the product ceases to be in any way unique and the system that produces it ceases to be a craft.

I have a friend who milks about 60 Holsteins twice a day. Every day a truck shows up at his dairy. The driver gets out, looks at the bacterial tests, loads the milk and drives away. At the end of the month my friend gets a check. He knows where the milk will be processed and approximately where it will be marketed, but that's about all he knows about the fate of his milk. People who buy his milk or the cheese made from it never know that he produced it. And he has no idea who bought his milk, or what they thought of it. He is a cog in a wheel that turns to the tune of commodity dairy. The check he gets at the end of each month is the only measure of the value of his work. And the factors that determine the size of that check are pretty much independent of anything that he does to influence the quality of the milk, as long as it conforms to bacterial and butterfat standards. The only way that he can increase the size of the check he receives is by increasing the quantity of milk he produces. I don't deny him the path that he has chosen. If it works for him, I am fine with it. However, I wish that his craftsmanship could be appreciated by those who drink the milk he produces or who eat the cheese made from his milk. I wish that he could know their satisfaction, their pleasure, with what he produces. I know it would make him happy.

I have another friend who, with his father, also owns a dairy. A few years ago they invested in pasteurization equipment. By pasteurizing the milk from their cows they can sell their products directly to the public. My friend's father runs the farm side of the operation and he runs the creamery. They estimated the lowest price that they could accept for their milk and still ensure the viability of the farm. The viability of the creamery, they decided, would always be secondary to that of the farm. Following specific guidance from Cornell University, the quality of the milk produced at this creamery was determined to be the highest in New York State in 2010. In addition to milk, the creamery produces heavy cream, half-and-half, ice cream, and (incredible) chocolate milk. Every morning, the younger partner gets into a truck and delivers milk to some of his customers. They know him. So do many of the customers who stop into the creamery to buy five or six gallons of chocolate milk. (Really, I saw somebody do this.) Folks who buy this milk at farmers' markets and supermarkets throughout the region may not know him but they do know the farm, and they seek out its milk because they know its quality. They seek it out because it is local and to many people that matters. They seek it out because they know who made it. That milk *is not* a commodity. There is too much information attached to it. There is too much pride and craftsmanship attached to it. There is too much that distinguishes this milk from all other milk — it is New York State's best milk — for it to fit the definition of a commodity. And that's the difference. If the farmer agrees to be part of the commodity system, his product will be a commodity. If the farmer opts out, his product will be judged by its quality and the skill that went in to producing it.

Of course, not every farmer can get a loan for the infrastructure, such as pasteurization equipment, needed to break free of the industrial system. Many are in too deep to even try. Others are doing well in the commodity system, and others are at least getting by. Who's going to spend money and accept the risks of trying something new when they are already making a living? Certainly, it's easy for me to sit here and write about jumping ship, when I know I'll never have to do it. It takes courage to add risk to an already risky business. That might be easier for a young guy on his own or with the backing of some family money. But it's a different story when you're feeding a family, when you're carrying all the debt you can handle and then some, and when you are at least breaking even. I am absolutely not implying that those in commodity farming are less sincere about what they are doing or less skilled than those who have opted to market their products directly to the consumer. To be a farmer is an act of courage and stamina regardless of what system one works in. Every single farmer is an artisan. In my opinion, however, the industrial system of agriculture shows respect for neither the craft nor the craftsman. Otherwise the product would not be relegated to the level of a commodity. My hope is that the public will come to value the craft of farming and recognize that all food is not the same. By doing so, one simultaneously rejects the central premise of industrial agriculture and gives meaning to the work of the one percent of us who produce food for the other 99 percent. My hope is that as the direct marketing of food becomes more feasible farmers everywhere will embrace it. Slow Food provides the simple logic that local, consumer-based food systems have value. Slow Money provides the implementation strategy.

PAM KLEPPEL

As we search for a less extractive and polluting economic order, so that we may fit agriculture into the economy of a sustainable culture, community becomes the locus and metaphor for both agriculture and culture.

— Wes Jackson, *Becoming Native to This Place*

12

CSA

I HAVE MENTIONED COMMUNITY SUPPORTED AGRICULTURE many times throughout this book, but the CSA concept is so important that it deserves a chapter of its own. The CSA model is transformative. It changes the way farmers farm, the way markets function, and the way consumers relate to the production chain.

The change has to do with risk and how it is distributed. As Wendell Berry pointed out, debt is the farmer's worst enemy. Debt, when added to all the ways that nature conspires to make a farmer's life uncertain, can push the farm off the cliff. The CSA model allows the farmer to manage debt. The CSA model also connects the consumer, not simply to farming, but to an actual farm. A connection to *farming* implies that one appreciates the craft of farming and the fact that as a result of the way it is produced food can be safe, nutritious, and flavorful.

All of this is very good. But one's connection to a *farm* is different. A farm is not a concept — it's a place. It is somewhere you can go. A farm is populated by the people who produce food and a community of consumers who depend on the farm for food. Through the CSA you, as a consumer, can become part of that place — that farm. You may even decide to participate in the production process. Or... you can simply show up each week and pick up your share. As a CSA shareholder you share in the harvest and in the risk. The CSA transforms the consumer into a stakeholder — not in agriculture, but in a farm... a *real* farm. That makes a difference in the way you relate to your food and its production. And it makes a huge difference to the farmer and to the ability of the farmer to manage risk.

I often open seminars and lectures by noting that "If you've eaten recently, you've participated in agriculture." We all have a relationship with agriculture, simply because we eat. Most of us, however, don't give much thought to our relationship with agriculture. The CSA changes that. It makes our relationship with agriculture personal and intimate. "Customer Service" is no longer an 800 number on a bag of vegetables. Customer Service is Ben and Lindsey, or Slim and Cathy, or Thomas and Liz... real people, with names, who produce food for the shareholders in their CSAs.

The CSA model grew out of the European anthroposophy movement founded by Austrian philosopher, Rudolf Steiner in the early part of the 20th Century. Anthroposophy combines science and spirituality. In agriculture, that translates into a deep respect for the soil — what today would be considered organic and, to a large extent, sustainable farming — and the biodynamic method. Biodynamic agriculture takes organic farming to another level. Unlike the conventional organic farmer, who may or may not use organic fertilizers and pesticides, the biodynamic farmer uses only composts and manures produced on the farm to promote the growth of crops and grass. Oh, and something else — biodynamic farmers use "preparations." These are prescriptive combinations of animal and plant materials, along with some soil, fermented in small pots and then buried in particular locations on the farm. From the preparation, biodynamic theory holds, the soil is cleansed and made healthy — that's the spiritual side of biodynamic farming. Whether biodynamics has reached the point at which science and spirituality converge, or whether it's just so much hokus pokus, is not for me to say. I would simply point out that there are an awful lot of really good and successful farmers who practice biodynamic farming. That's good enough for me.

The "community" farm concept took hold in Europe in the 1960s. Community farms were organized around relationships between food producers and consumers. The farm was a community supported by the work of the farmers — who more often than not practiced organic or biodynamic agriculture — and by the patronage of consumer-investors in the farm. The concept spread to North America in the 1980s with the founding of the Temple-Wilton Farm in Wilton, New Hampshire and the Indian-Line Farm in Egremont, Massachusetts. Temple-Wilton was established by a German biodynamic farmer, Trauger Groh, and his American collaborators. Indian-Line was established by Robyn Van En and Swiss bicycle builder Jan VanderTuin. Van En is credited with promoting the CSA movement worldwide and personally starting more than a thousand CSAs.

CSAs take many forms but the thing they all have in common is that prior to the planting season consumers purchase a portion — a share — of the harvest, for about $300 to $500. Some farmers sell half shares. Some sell shares for additional products such as bread, fruit, meat, or poultry. One year Pam and I bought a cut-flower share in addition to our vegetable share. But consumers always pay for the harvest before the season, thereby sharing with the farmer both the harvest and the chance that some, or all, of the crop will be lost. Each week during the season, shareholders receive their portion of the several kinds of produce harvested on "pick-up" day. Many CSAs require pick-up at the farm. Others deliver the shareholders' boxes of produce to specified pick-up locations — shareholders' porches, hospitals, universities, and so on. But if disaster strikes either the farm or the farmer, the "community" — the farmer and the shareholders together — share in the loss.

The CSA model is evolving rapidly, and in dramatic ways. In the early days shareholders were provided with a box of vegetables, plain and simple. Increasingly, however, farmers are exploring ways to address shareholder preferences — heirloom tomatoes are popular; kolrabi should be dispensed sparingly. These days many CSAs allow their customers to "load" their own boxes — a set weight or volume of each kind of produce being allotted — and to exchange one or more unwanted items at the "swap" box (deposit what you don't want, take what you do).

Efforts to respond to customer preferences have spawned a "free choice" CSA model. The shareholders in Cathy and Slim Newcomb's CSA pick up their weekly shares at the farm. Each share is composed of about $33 worth of vegetables. Customers shop in the rustic little "farm market" next to Cathy's hoop-house nursery (customers are free to browse) and can choose whichever vegetables they want. Prices are marked; the total cost of the items selected must be $33 — more or less. All-in-all, it is not a bad way to shop. As a topper, customers can add a jar of pickled tomatoes, an artisanal vinegar, or one of Cathy's incredible jams (Gram's Jams).

One problem that arises is that shareholders can do only part of their shopping through the CSA. Perhaps they frequent a farmers' market as well. And, ultimately, they will have to visit the supermarket for paper goods, household cleaners, and so forth. The question then becomes how much time will they spend shopping each week? And really, if they are patronizing a farmers' market, why do they need a CSA? Clearly, while the value of the CSA model for the farmer is obvious, busy nine-to-fivers may not find it particularly convenient or relaxing to add a farm stop to an already crowded schedule.

CSAs that deliver to workplaces help to resolve that problem. At the university where I work, CSA shareholders stop at the library after work on Friday to pick up their shares on the way to the parking lot. An iteration of that, which Pam and I have found useful during the winter, is a service that buys gleanings from farmers in the Hudson Valley throughout the fall, well after our CSA has quit, and frozen local vegetables during the winter. These, along with whatever root vegetables can be scavenged and the ever-present micro-greens, are put together to create a "weekly bag" of vegetables which can be purchased either on a pay-as-you-go or on a contractual basis. The weekly bags of vegetables, along with cheese and bread shares for those who are interested, are dropped off at the law school where Pam works and at several other institutions in our area.

The breadth of the CSA model is constantly being tested. And nowhere is this truer than at Essex Farm, near Lake Champlain on New York's border with Vermont, where Mark and Kristin Kimball are experimenting with a year-round, free choice CSA model. Mark began farming after graduating from Swarthmore. Harvard grad, Kristen left a career as a travel writer to join Mark on his farm. But, she hasn't left her career as a writer completely; Kristin's first book, *The Dirty Life,* is a wonderful memoir of the thirty-something couple's experiences farming. The Kimballs started their whole-diet, free choice CSA with the knowledge that their farm allowed them to produce great food for their family all year long. It followed that if they shared their harvest, many other people could eat well too, and they could pay the mortgage on the 500+ acres they were farming. For about $3,000 per person per year (about what you'd pay at the supermarket) each member of the shareholder family can select from beef, chicken, pork, dairy products, myriad organically grown vegetables, and several kinds of berries. Children under 12 eat free. The Kimballs don't do soda or Twinkies®.

The CSA is an experiment-in-progress. The future of the CSA model might well be a convergence with farmers' markets. Tilldale Farm, which sells their pastured beef, pork, and chicken at the farmers' market in Delmar, New York, has a meat CSA. Shareholders pick up at the farmers' market. Mariaville Farm offers a similar deal to customers who shop at the Troy Farmers' Market and the Schenectady Green Market, in upstate New York. While there, CSA customers can fill most of their grocery needs — vegetables, bread, condiments, and so forth from other vendors. The Tilldale and Mariaville CSAs bring customers to the market, which vendors appreciate. And, the shareholders gain access to additional produce, perhaps obviating the need for a trip to the supermarket (unless, of course, they're low on paper goods).

However the model evolves, the future of sustainable farming is certain to be tied to CSA and similar tools that help farmers manage risk, while ensuring consumers the freshest, safest, most nutritious and tasty food available. The very phrase, community supported agriculture, identifies the vital, logical, and inexorable connection between the farmer and the community. How the CSA model evolves going forward is limited only by the creativity of the farmers and consumers who collaborate in the emergent agriculture.

Mark Schmidt

The local deal is the best for everyone… an apple is an apple; large… companies don't add anything to it except costs.

— Peter Ten Eyck, Indian Ladder Farms, Slingerlands, New York

13.

Scaling Up – How Local Farms Will Feed America

AGRICULTURE SCALES TO MARKETS. Small farms can meet the needs of local, consumer-based markets. Large farms often require commodity-based markets. A 50-acre farm can no more feed a city than a 5,000 acre farm can get its entire income from a few farmers' markets. The emerging paradigm in agriculture will require that the issue of scale be considered. The question is this: Is small scale, local agriculture a real alternative to the industrial system? Or stated another way: How can small "local" farms produce enough food to feed the American people?

Before going further, we must consider the issue of "localness." What do we mean by "local food" and why does it matter? I've touched on this issue several times in this book but, given the subject of this chapter, it seems worth going into a bit more detail.

Most definitions of local food focus on distance, which seems reasonable until we look at the data. Professor Christian Peters of Tufts University studies foodsheds. He and his colleagues define "local food" as being food produced within about 30 miles of some reference location (generally an urban center).[1] Meanwhile, Rich Pirog and his co-workers at Iowa State University observed that fruits and vegetables traveled an average of 56 miles to "local" farmers' markets.[2] Alisa Smith and J.B. MacKinnon decided to "eat locally" for a year and wrote a book about it.[3] They defined "local" food as food produced within 100 miles of one's residence. They discovered that supermarkets were unable to provide much food that met their definition of local, so they shopped at farmers' markets. The Association for the Advancement of Sustainability in Higher Education (AASHE) stretches the definition of local food to "within

250 miles of campus." The supermarket where I buy groceries has a local food section. Their website has photos of farmers who produce that food. However, when I asked a clerk at the supermarket what they considered local, she couldn't tell me. Overall, it seems that the distance that defines local food is largely arbitrary, dependent on personal or institutional expedience rather than some functional relationship between food and distance.

There does, however, seem to be an interesting connection between what people consider local and the way they think about the food sold at farmers' markets. Perhaps localness has more to do with farmers and markets than with distance. Perhaps the question ought to be "What do people get by shopping at a farmers' market that they can't get at, say, a supermarket?" At this point it cannot be stated with confidence that eating locally saves energy (though many people probably believe that it does). In Chapter 7 (Energy and the Future of Farming) I pointed out that a comparison of the energy costs and greenhouse gas emissions associated with transporting food long distances in a few large trucks versus those of transporting food short distances in many small trucks are fuzzy at best. Perhaps better answers to the questions posed above, however, have to do with information and nutrition. When I visit a supermarket, I am bombarded with labels about my food — cage free, free-range, organic, naturally raised, all natural, local — most of which tell me nothing about the food. Except for USDA certified organic and free-range, none of the others have any standard or enforceable definition.

When I purchase something, say a steak or a carton of raspberries, from a farmer at a farmers' market, I establish a direct connection to the processes used to create those products. That connection is the farmer who hands me the food. Compare the potential information content associated with food at a farmers' market with that available at a supermarket, where workers might not even know what the store considers "local" food. At the farmers' market, I can question the people who produced the product. I can find out how the producers define "cage-free," "pasture-raised," and "free-range." I usually can't do that in a supermarket. Now it is certainly possible that a farmer can lie, even though getting caught in a lie may be worse than not selling the product. I believe, however, that most of the people who sell food at farmers' markets are honest. (I am biased, of course, because I'm one of those people.) The reality of quality control in small-scale, local production systems was stated succinctly by Peter Ten Eyck, who owns Indian Ladder Farms, one of the two remaining orchards in Albany County, New York. "I need to keep six clipboards going for every gallon of cider I produce," Ten Eyck complained at a recent US Senate town hall meeting dealing with government regulation of agriculture. "What's the point? If somebody gets sick because of my cider, I'm finished, clipboards or not." The farmer at the market has a lot at stake. The people who shop at farmers' markets want to talk about how the food they're buying is produced. Farmers who sell at markets know this and make themselves available for questions. More importantly, farmers who sell at farmers' markets know that if someone gets sick and the illness is traceable to their food, they are finished. Not only that, but the market is finished, or at least has a great deal of damage control to do. So if "local" implies a direct connection between producer and consumer, then two things are probably true. First, the food is safe. Second, the information content about the food — how it was produced — will be high, relative to markets in the industrial food system populated by middlemen who've often had little or nothing to do with producing the food they are selling.

There is something else about localness that *is* associated with the distance the food travels. It has to do with quality. The freshness, nutritional value, and taste of food vary as a function of the length of time between the moment it is harvested and the moment it is eaten. Long before "local" and "farm-to-table" became chic, chefs at the nation's best restaurants shopped at farmers' and open air markets. Busy chefs would not make that effort unless they were convinced that fresh, local ingredients made a difference. When I was a graduate student, I learned to make artificial growth media in which to raise the algae that I was studying. The media required a mixture of vitamins. After preparing the vitamin mixture I would freeze it because vitamins are "heat labile." That means that as the temperature rises the rate of degradation of vitamins increases. In the freezer at -4 degrees Fahrenheit the vitamin mix might last two or three months. In a refrigerator at 40 to 50 degrees it could go about a month, and in the incubator in which I grew my algae at 68 degrees that vitamin mix would last two weeks at best. Refrigeration slows but does not arrest biological or chemical processes. We all know this because we have all experienced milk souring when it is left in a refrigerator for more than a week or so and the take-out that we forgot about in the back of the fridge, that will have grown "hair" over the same period of time. Scientists at the University of California, Davis determined that fruits and vegetables may lose nearly 50 percent of their volatile polyphenols (antioxidants that also impart flavor) and vitamins during transport.[4] And research in Spain has reported that, after seven days of cold storage (to simulate transport) and three days at 60 degrees Fahrenheit to simulate the grocery shelf, broccoli lost 50-80 percent of its vitamin C and the volatile organics that impart flavor and provide nutritional benefits.[5] A friend of mine, Jim Kittredge, who was a chemist at the University of Southern California and also an avid gardener, used to take a pot of boiling water along with him when he harvested his corn. He shucked the ears on the spot and tossed them into the boiling water. He explained that the moment the corn is picked the sugars in the kernels begin changing to starch. If you got the corn into the boiling water immediately the sugars remained intact. Jim's sweet corn was amazing! Both the anecdotal and scientific evidence point in the same direction —chefs know what they are doing. Vegetables at the farmers' market, picked at five in the morning, are chemically different from vegetables picked two weeks ago while under-ripe on a farm in California, packed onto a truck and transported 3,000 miles to the East Coast, then unpacked into the produce section of the supermarket and left there for as long as a week. And the vegetables in the farmers' market and the supermarket are both labeled "fresh."

Taking all of this together, the attributes of space and time only give meaning to the definition of local food within the context of information and quality. The functional value of local food is determined by the customers' ability to ask the producer questions about the production process, and by the freshness, i.e. the nutritional and flavor characteristics, of the food. There is no spatial threshold beyond which food is no longer local except as determined by the extent to which the separation of producer and consumer limits communication between them or is detrimental to the flavor and nutritional quality of the food. Food is local when we have the ability to ascertain how it is produced. Food is local when its nutritional value, flavor, and freshness are optimal. These criteria may be best met within a geographic distance that is correlated with one's ability to speak directly to the food producer.

Geographic localness also creates a connection with the financial viability of farms in one's community. Throughout this book I have emphasized the fact that if a farm is to succeed, it must be profitable. To ensure both the quantitative and qualitative values that farms provide, farm incomes must be supported by the market. Farms that sell their products directly to consumers, rather than to wholesalers or distributors (i.e. through commodities markets) depend on those consumers to ensure their profitability. Agricultural viability created by consumers provides returns to the community many times over. The economic multiplier effect of local agriculture, i.e. the number of dollars generated for every dollar spent, is, conservatively, between two and three. That is, for every dollar spent purchasing food from a farm in one's community, the community will realize $2–$3 in economic development. The implications of the multiplier effect are intriguing. If, for instance, each of the 18 million people who live in New York State spent just $10 on locally produced food each week, agriculture would stimulate between $62 and $94 million dollars in economic development annually. That same equation applied to the entire United States would yield an economic stimulus between $326 and $450 billion. By whatever spatial standard one uses, most (about 68 percent) of the money spent within one's community stays there.[6]

The safety and security associated with local food, and the economic implications of eating locally, are prerequisites to sustainability in the farm-to-market equation, but the question remains: Can local farms feed the nation? Many would say no, we cannot feed 314 million people by sourcing our food locally. It is simply more efficient to "assemble" our food in factory farms and then spend the energy to get it where it needs to go. How would all that local food get to population centers such as New York, Boston, and Los Angeles? How many produce-laden pick-up trucks would need to converge on New York City each week to feed its nine million residents? And how could a university that serves 30,000 meals a week even think about local sourcing?

I have participated in finding an answer to the last question, and that answer implies that there is a rational solution to each of these questions. The answer comes in two parts. First, eating locally should not be thought of as an "all or none" process. We don't have to feed every American a complete diet from a farmers' market in order to consider the process successful. If by, say, 2025 every American home and institution can conveniently fill half of its food budget from local sources, we'll be doing well, and so will American farmers and our economy.

The second part of the answer is that we will not succeed at local sourcing on a large scale using current food production and distribution models. As the age of cheap oil comes to an end (and, in case you haven't noticed, it already has) production and distribution systems are going to change, whether we like it or not. They already are changing. As the cost of transporting food (and therefore the cost of food at the market) rises, local production and distribution become increasingly appealing.

Localization of agriculture cannot occur, however, with a food production system that is subdivided into artificially-maintained regions, as it has been for the past 70 years. For example, more than 70 percent of the nation's fruits and vegetables come from California's Central Valley. Production in the valley is maintained through a completely artificial and heavily subsidized irrigation network that has drained lakes and severely modified the flows of western rivers. This situation is not sustainable indefinitely.

Although some regions of the country are uniquely suited to producing certain kinds of crops — oranges in Florida and California, for instance — much of the nation, particularly its temperate zones, are capable of producing an enormous diversity of fruits, vegetables, and meats. Greenhouses, hydroponics, and cold storage technologies make fruits and vegetables available to consumers year-round — I eat New York apples in July and New York tomatoes in February. The shifting agricultural paradigm requires that every region of the country produce the diversity of products that it is capable of producing — that monoculture and exclusive production zones be eliminated to the extent possible. When every region in the country produces all of the kinds of agricultural products that it is capable of producing, the capacity to feed ourselves locally increases.

While the transition to a locally-based agriculture may seem like "pie-in-the-sky" today, it is in fact the way agriculture has worked for most of the past ten-thousand years. Only within the past 100 years (one percent of agriculture's time line) have cheap fossil fuels (and the development of refrigeration) made transporting food long distances seem reasonable. As cheap oil becomes a thing of the past, so does the logic of industrial monoculture. Ultimately, agriculture needs to change, and it is.

Part of that change is measurable as an increase in the procurement of locally produced food at institutions such as colleges. I have been involved in that process at the State University of New York at Albany, where I am a professor. In the summer of 2008, I spent hours on the telephone with representatives of the parent company of the food service corporation that provided all of the university's dining services. I was collaborating with Kathleen Harris of the Northeast Livestock Processing Service Company (NELPSC), a consortium of New York farmers, whose mission is to provide sustainably-produced local meat to institutional markets. We were trying to get the food service to buy meat — particularly ground beef — from locally raised, grass-fed cows. The problem was that the food service required a $5 million bond to protect the company against lawsuits from people who got sick on the meat. In addition, they required an inspection of the packing plant used by NELPSC by an outside consultant (called a third-party inspection). The inspection criteria, created by the food service, could not be met by a small processor even though the processor met all USDA food safety and handling requirements.

On a hot Friday afternoon in July, I was on the phone with a receptionist at the corporate headquarters of the food service, talking about the importance of getting local food into my institution. Finally, the receptionist transferred me to a vice president who, as it happened, had grown up on a farm. She was sensitive to the values and needs of small-scale agriculture and, after about half an hour, was convinced that grass fed animals simply could not produce acidophilic *E. coli* O157:H7 — the pH of their rumens would just not support it. Further, I argued, the processing of livestock at a small packing facility is probably more closely observed by USDA inspectors than it is at large plants that process 300 animals an hour. She agreed. "I'm going to ramp this up," she said. Two weeks later the bond required by the distributor was lowered to $2 million. Soon after that, the company's packing plant inspection policies were reviewed and, with help from NELPSC, revised so that a small facility could pass the third-party inspection. Three months later, the University at Albany purchased its first 2,500 pounds of grass fed ground beef from NELPSC.

In April 2009, I convened a Local Foods Conference at my university. Students, faculty, and food service personnel from around the region participated. At the time, eight percent of the food served in our university's dining halls and eateries was locally sourced. I suggested that we should aim to achieve about 50 percent local sourcing by 2015. The director of the food service took the podium and remarked that we would be lucky if we could achieve 20 percent local sourcing in 10 years. My students and I really couldn't accept that — not from a contractor. As I write this (August 2012), the university locally sources 30 percent of the food it serves. The motivation that led to the dramatic increase in local sourcing by the food service came partially from the faculty and some key administrators. But mostly it came from students, the university's customers, to whom both the university administration and the food service are very responsive.

To increase the students' motivation to encourage the university to make a serious commitment to the local sourcing of food, it was necessary to provide information to them about the differences between locally and conventionally produced food. To do that, I offered the 700 students in my freshman biology class a little extra credit if they would attend a teach-in on local food, on a Saturday. One of my graduate students suggested that we pass around a petition during the teach-in, requesting that the university purchase more food from local sources — students were under no obligation to sign. On the day of the teach-in, the lecture hall was packed. I made the case for local sourcing, explaining the health and safety implications, and the environmental and economic consequences of different food systems. I also explained that there is no problem getting local, greenhouse, or hothouse-grown vegetables in mid-winter in upstate New York, and that locally produced food doesn't necessarily cost more than food trucked 3,000 miles from California. I explained that livestock fed grass are safer than cattle finished in feedlots, fed antibiotic-laced grain, and sporting steroid implants to "help" them grow faster. By the time I finished my lecture, the students were energized and we had 600 signatures on the petition. A delegation of undergraduate and graduate students took the petition to the office of the Vice President for Business and Finance, and the student campaign to source 50 percent of the university's food locally by 2015 — "50 by '15" — was underway.

The local food movement on campus takes several forms. Each element of the program has a distinct impact, economically and socially. One of the most significant elements is a small farm-stand operated throughout the fall at our Campus Center (the same thing as a student union) by Jim Abbruzzese, owner of Altamont Orchards. Every Thursday, Abbruzzese sets up a table full of produce from the orchard. Students get to sample apples, cider, and cider donuts — an upstate New York delicacy. The students love the food and they love having Jim on campus. For the first time in their lives (for most of them) they can take an apple from the hand of the person who grew it. They can talk to him about his apples and tell him that those donuts are AWESOME. They can tell him about the screwy professor who assigned tasting a cider donut as homework. And he will grin in agreement. As at any farmers' market, the people who produce food and those who consume it connect — a bond is formed that gets stronger with time. Now Jim isn't going to retire any time soon on what he makes at his university farm stand, but I know he realizes that what he's doing is important. He does it because it's the right thing to do. It is one thing to crate 50 bushels of apples and send them off to a nameless, faceless distributor in Cincinnati or Chicago, but it's quite another to hand an apple to a kid from the Bronx who has never even seen a farmer before, much less taken an apple from the hand of the person who grew it.

Progress in the campus local food movement has been made in fits and starts. Student and faculty support has been critical to ensuring that the university administration continues to aggressively pursue a policy of local sourcing. However, the expertise to create and implement that policy was lacking. In 2011, the university hired Stephen Pearse — a veteran of the food service industry — as Executive Director of University Auxiliary Services (UAS, the non-profit corporation that holds the university's external service contracts). Pearse understood that the key to local sourcing was to find a person who understood New York's agricultural community and had the connections and experience to create a model of local procurement that could meet the university's needs and the goals of its local food program. He looked to a pair of old friends, Dan and Vickie Purdy, of Purdy & Sons, for that expertise. Dan Purdy is a third generation butcher in upstate New York. With his wife, Vickie, he has developed a clever model of local food procurement for the university. First, Dan and Vickie knew that small farms could not handle the food needs of a large university, and that most large farms in the region were already tied to the commodities markets, and would not be interested. Mid-sized farms, however, were well suited to the university-scale. Dan offered farmers 15 percent above commodity pricing, and the potential for long term investment, to produce food for the University at Albany. Farmers immediately recognized this as a very good deal. Pearse determined that, with a little modification, the costs associated with local sourcing could be fit into the UAS food budget. Local sourcing rose to 30 percent during 2012-2013 academic year. I recently asked Steve, "Will our university be 50 percent locally sourced by 2015?" "We can do that," he replied.

The lesson that I took away from the University at Albany's efforts to bring locally and sustainably produced food to our campus is that most American colleges and universities (and other institutions) *can* reduce their dependence on industrially produced food and food transported long distances. They can readily obtain a substantial portion of their dining needs locally. The amount of food sourced locally and the rate at which an institution makes the transition to local sourcing will depend on the commitment and passion that individuals, particularly students, bring to the effort. Initially, individual institutions will come up with their own processes for obtaining their food locally. As each finds out "what works," they will begin to share and to integrate successful strategies from the emerging network of institutions using some iteration of the emergent suite of local sourcing models. The process gets easier as the local foodshed develops and the new system becomes the institutional norm. As students graduate and leave the college, they will take with them an understanding of the value of a local food economy. They will carry what they have learned into their personal purchasing and eating spaces, as well as into the institutional environments where they work and where their children are educated.

As the model for accomplishing the conversion to local sourcing develops and matures, the pace will invariably quicken. A decade ago, few colleges were interested in local procurement. Today, the entire State University of New York, the largest public university in the United States, has committed to local food procurement. Aside from the health, safety, and gastronomic values of local food to those who eat it, the economic benefit realized by the State of New York as a result of local food sourcing by its State University system will be on the order of hundreds of millions of dollars annually!

The local food movement is not confined to colleges and universities. Primary and secondary, public and private schools are deeply involved, as well. Every state in the union has at least one "Farm-to-School" program. Certain states — including Minnesota, Wisconsin, and New York — have more than 100.[7] While each program is independently managed, many involve students in the growing of food, and even preparing the meals served in school cafeterias. Hospitals and even prisons are involved with the local food and gardening movements. States such as Florida and South Carolina consider farms in prisons as a means of reducing the high cost of maintaining prison populations.[8] At New York's notorious Rikers Island, the Horticultural Society of New York runs the Green House Project, which rewards good behavior by allowing inmates to work in the garden.[9] Prisoners learn to grow horticultural products as well as food. According to the Warden at Rikers, "These types of programs not only enhance the environment by increasing the green:asphalt ratio, but growing food near prison sites improves the nutritional intake of the inmates, as well as trains them for green jobs when they get out..." The recidivism rate among ex-convicts who took part in the Green House Project while incarcerated is about a quarter of the national average. Similar results have been obtained in Lawrence, Massachusetts, where some 350 inmates in the Essex County Correctional System take part in a farming program.[10] Recidivism among these inmates is 20 percent lower than at the county's main jail.

Using the spatial analysis tools of a geographic information system, Dr. Christian Peters of Tufts University and his colleagues at Cornell showed that on average, the people of New York State can meet 34 percent of their food needs with food produced within a 30 mile radius (I mentioned this study above). When New York City is removed from the analysis, however, the food needs of the rest of the state's population are easily met locally. But we can't simply delete New York City, with half of the state's 18 million residents, and claim that New York State can easily meet its food needs from local sources. Perhaps a 30-mile standard is a bit harsh. The AASHE 250-mile definition of "local" may be more reasonable for urban centers whose fringe areas have been consumed by suburban sprawl. Under the more "institutionally-scaled" AASHE definition many cities with prolific "green market" programs, food cooperatives, and farmers' markets would be considered capable of feeding their populations "locally" produced food. But this is where definitions of local can become a bit misleading, and where concerns about the impacts of transporting food into cities become valid.

Perhaps the most important alternative for locavores in urban centers is urban farming. From New York to Los Angeles urban gardens and farms are sprouting in alleys, on roof tops and vacant lots.[11] In Detroit, inner city farmers are turning urban food deserts into gardens, producing fresh vegetables and some meat, feeding people whose diets were a national disgrace beautiful, fresh, nutritious food grown blocks, not miles, from home. Former collegiate basketball star and MacArthur Genius Grant winner, Will Allen, has turned a bankrupt nursery in Milwaukee into an icon of the emerging urban farming movement.[12] Allen and his contemporaries have shown that, by learning to produce food, one gains self-respect, personal health, and peace of mind. Urban farming is going vertical as well. City roofs from San Francisco to the Bronx are covered with fruits, vegetables, and flowers. In Singapore, engineer Jack Ng has founded Sky Greens, an ambitious project in which vegetables are grown in vertically rotating trays.[13] At the bottom of the system is a pool of nutrient rich water. Once bathed in the nutrients, the trays are lifted slowly upward into the light, by a gravity powered water wheel. The system produces 10 times more food per square foot than conventional farming at one third of the energy costs.

While trucks full of food from the hinterlands will continue to roll into the lucrative green markets of the nation's cities, the real story, I submit, will come from within. Inner city revitalization through the process of growing food may, at the end of the day, be the most significant achievement of the emergent agriculture.

Part IV
Conclusion

G. Kleppel

Pam Kleppel

Cultivators of the earth are the most valuable citizens. They are the most vigorous, the most independent, the most virtuous, and they are tied to their country and wedded to its liberty and interests by the most lasting bonds.

— Thomas Jefferson

14

The Emergent Agriculture

It is nearly noon on a warm Monday in July. I'm sitting at my writing desk in the kitchen, the place where I began this book. As I have written, I have learned. My understanding of agriculture and sustainability has deepened. I see more clearly than ever the positive impact that the emerging paradigm in agriculture is having on our nation's food systems. It is reviving the crafts of producing food and distributing it locally. It is creating a healthier, safer, more nutritious food supply. However, in very real ways the emergent agriculture is creating something larger. It is fostering an ethic of environmental stewardship, social justice, and animal husbandry, attributes ignored during the past six decades, when the extreme industrialization of food production sacrificed food safety, quality, and often ethical behavior at the altar of profitability. The emergent agriculture is returning humanness to the craft of food production. The emergent agriculture celebrates the ethic of land health[1] and those who embrace that ethic as they use the land in pursuit of their craft.

It is here, in my farm kitchen, where I began to truly understand how a new way of farming and a new breed of farmer are addressing the public's yearning for safe and ethically-produced food. The destabilizing commodity markets are being replaced, albeit a little at a time, by markets in which producers and consumers know each other and recognize the importance of each to the other and to the sustainability of the system. Realistic alternatives to volatile commodities markets are emerging as well. The explosive increase in farmers' markets (up 250 percent), CSAs, and other forms of direct marketing over the past 30 years illustrates the power and intensity of the change that is underway. The instability of monoculture is being replaced by stability-creating crop diversification. Decentralized food production systems, in which decisions are made by farmers in the field, are replacing the vertically stratified centralization of corporate agri-business and multinational conglomerates in which decisions are made in boardrooms, absent concern for the land, livestock, or co-workers. The emergent system allows the farmer to take their case — that what they do cannot be duplicated in the agro-industrial complex or in a laboratory — directly to the consumer. And when the case is made, consumers embrace the emergent system passionately. I know this, because I experience it every week at the farmers' market where I sell my products.

I have learned that food production becomes exceptionally sustainable when profitability ceases to be the singular mission of the farm, and becomes instead a part of a "triple bottom line." As important as profitability is, true success demands that we focus on soil health and the functionality of the farmscape ecosystems that "subsidize" the farm, as well as the quality of the lives of the farmers and even of people far away who they'll never know. Coincident with healthy soils and functional ecosystems is diversity — both biological and operational. Biologically diverse ecosystems provide buffers against disaster, a critical function in an increasingly unstable environment. Operational and economic diversity — through intercropping, inter-grazing, multi-use, and diversified marketing — enhance financial stability, even as global commodity markets teeter on the brink of collapse or rise and fall like an economy with bi-polar disorder. The third bottom line is ethics. It is the bottom line that emerges when one seeks to answer the simple question, "What's the right thing to do?" When every aspect of the farm operation, including the relationship of the farm to the market, is considered from the perspective of ethical behavior, then both environmental stewardship and financial profitability are emergent. Humane treatment of land and livestock, respect for the skills of farmworkers, and honesty in food production and marketing, are essential to sustainability in agriculture. These are the elements of farming in modern times — of the emergent agriculture.

The role of the farmer includes that of caretaker — one who ensures that the modifications to nature that agriculture demands do not weaken the overall system. The underlying premise of industrial farming is that through technology we can supersede nature. The modern, sustainable

MARY ELLEN MALLIA

farmer knows that we can't, and that we will hurt ourselves if we try. The modern farmer understands what is required of him or her — their responsibilities to the system — not because of the consequences, but because "it's the right thing to do." The modern farmer knows that to be viable and to prevail in a world full of risks and uncertainty, relationships based on respect and recognized responsibility must be forged. Other relationships must be severed. The relationship to commodity markets, to synthetic chemical inputs, and the commitment of massive amounts of petroleum to every part of the operation must be re-assessed. The relationship to science and technology as irrefutable goods must be infused with a personal relationship to the systems to which the science and technology are to be applied. To paraphrase Wendell Berry: animal science must become animal husbandry; soil science must become stewardship. Science and technology, while essential to progress, are only truly representative of progress when accompanied by the farmer's concern for all of the systems associated with the farm — the soil, wild nature, the livestock and crops, the business and the consumers that the farm serves, as well as other farmers. Most critical are the relationships between farmer and consumer. Without support from consumers the farm will die. Without consumer understanding of what goes into sustainable production, there can be no support. When the importance of connectivity and relationships is appreciated a stabilizing farm-community and, ultimately, a network of communities emerges. With few exceptions people are hungry for these relationships. They are sustained by these relationships and they prosper from them. The emergent agriculture celebrates ethical relationships between farmers and consumers, farmers and livestock, farmers and the land.

My kitchen is warm today, but a cool breeze blows in from the open east-facing window as I contemplate the final chapter of this book. In reality, I will probably not write the final chapter on the emergent agriculture — at least not for a long time. This is just the beginning. How this new, sustainable approach to farming plays out will not be known for years — very likely the process will continue to evolve, a clear endpoint never emerging.

Or perhaps as the movement grows and actually challenges the industrial system, the corporate juggernaut will begin churning. Allies in government, in the courts, and in the media will be mobilized and the insurgency will be crushed like a weed beneath the boot of a giant. Alternatively, the emerging tide of sustainable agriculture may sweep away the old and the stale, leaving in its wake a system of food production and consumption that does not deplete resources, degrade ecosystems, abandon ethics, commodify the life support system, or dehumanize the people who produce our food.

To be sure, the revolution is further along in certain places than others, but it is gaining momentum everywhere. Farmers' markets and other forms of direct marketing represent, nationally, the fastest growing sector in agriculture. Increasing numbers of consumers are seeking humanely produced food. Small-scale organic production, the local food movement, and farm-to-table culinary venues are proliferating.

Is it a fad? Time will tell, but I would bet it's not. Decades of work by scientists, essayists, investigative journalists, and documentary film makers, not to mention farmers, are beginning to bear fruit. With public exposure to the accumulating body of knowledge about how food is produced and marketed has come greater scrutiny of the system and greater concern about its safety and quality. Emerging from the increasing public dissatisfaction with the currently dominant food system is a cadre of farmers and producers who believe in the ideals expressed in these pages — that food is not a commodity; that food needs to be produced ethically; that the quality of our food, rather than just the amount produced, matters. I have argued throughout this book that as the emerging market, composed of informed consumers, converges with the emerging approaches to farming by people who respect the principles of sustainability, the paradigm of food production and distribution in America will shift. This is not a fad. It is the future of our food system.

The market share of the emergent agriculture is miniscule. Yet, it is changing the nature of the American food system in clear, measurable, and positive ways. Ten years ago, most four star restaurants would not disclose the sources of their produce, meats, and breads. Few customers cared about where their food came from. Today, the names of the farms that supply these restaurants are clearly displayed on their menus, along with such adjectives as grass-fed, heirloom, local, and sustainably-raised. I have been in restaurants in Philadelphia that claim to own their own farms. People care. And that's what's changing the system. To be sure, the Sysco truck is not sitting idle. The system is not going to change overnight, but it is changing.

The fact is that farmers just can't keep farming the way they have for the past 60 or 70 years — running environmental and economic deficits and ignoring the ethical questions that confront us every time we enter a supermarket, or a barn, or look out on 3,000 acres of genetically modified soy. People want the products that sustainable agriculture has to offer. People want good tasting, clean, nutritious, safe food. People want to support their local economies. People want to know that what they eat was produced humanely. People want farms in their neighborhoods, regions, states, and nation. People want to know that the iconic perception of American agriculture actually describes the system that produces their food.

If I had lived in Boston in 1770 and written a book suggesting that a revolution was underway that would create a sea-change in the American colonies, a few activists might have agreed but the majority of colonists at that time were happy being British subjects and could not have conceived of such radical change. Six years later that change would have seemed considerably more reasonable, and eleven years later they would have no longer been British subjects, but citizens of the United States.

The idea that a revolution is underway in the food system may seem no more rational than the idea that the American Revolution was underway in 1770. But the signs are clear — from the growing interest in sustainable agriculture, to the explosive growth of direct marketing, to the local-food and farm-to-table movements in food service. A revolution in agriculture is clearly underway.

Agriculture is no stranger to revolution. The very concept of agriculture was revolutionary ten thousand years ago. And while agriculture spread slowly from a few "local centers" throughout the world, ultimately it became emergent. The more recent Industrial Revolution led to a change in agriculture that initially improved the ability of society to feed itself, but that revolutionary system has devolved into something that is costly and destructive of the ecological and social systems upon which society depends. It has to change. The emerging breed of farmers, who seek a more benign relationship with the earth and a more humane approach to producing food, aren't looking to return to the good old days. Rather, they are making progress toward the better *new* days. They recognize that farms are natural systems. They understand that farms are constrained by the same rules that govern all natural systems and that breaking these rules has consequences. Farmers understand that agriculture depends on climate and energy systems and that these systems have become increasingly unstable and unpredictable, in part because of the way we produce food. The new breed of farmer is exploring systems of farming that reduce risk while not contributing to further environmental or economic destabilization.

Pam and I, like many of the farmers I've described in this book, are cognizant of our place in the sustainable future of agriculture — of the fact that it's not just what we produce but how we produce it that matters. We understand and take pride in the fact that the impact we have on the food system, as small as it is, matters. The simple daily routine of the farm tends to conceal the broader significance of agriculture. But as we finish dinner each evening and, instead of heading to the den to watch the news, we head to the barn to begin our evening chores, we are reminded of the commitment we have made to producing food — food for people, many of whom have become our friends through our relationship as producers and consumers. As we pass the solar panels, we are reminded why we've made this commitment. Pam begins to sweep the barns and I climb onto the tractor and head out to turn the compost. Soon Tory, my border collie, and I will bring the sheep back to the barnyard, and Pam and I will check the lambs one last time. We'll button up the chickens in their coops and give the dogs one last run. We'll be reinforced by our conviction that what we're doing is the "right thing" and, as always, we'll be grateful for our lives here at Longfield Farm.

MARK SCHMIDT.

Endnotes

Introduction — On the Cusp of a Revolution

1 Between 1763 and 1765 Parliament sought greater enforcement of sugar and molasses tariffs and passed the Currency Act, prohibiting the printing of paper currency in the colonies; the Stamp Act; the Quartering Act, requiring that the colonies provide barracks for British soldiers; the Townsend Acts, taxing glass, lead, paint, paper and tea; and the Declaratory Act, allowing Parliament to pass laws that were binding on its colonies.

2 Lewin, R. 1999. *Complexity: Life at the Edge of Chaos.* University of Chicago Press.

3 Agricultural Marketing Service, USDA. As of December, 2013: tinyurl.com/llg3cxb or ams.gov

4 Geneseo Food Research, 2010. History of Community Supported Agriculture. wiki.geneseo.edu/display/food/History+of+Community+Supported+Agriculture

1. A New Approach to Agriculture

1 From the film *The Greenhorns* produced by Severine von Tscharner Fleming (2011).

2 Ibid.

3 In his classic works, *An Agricultural Testament* (1943, Oxford University Press) and *The Soil and Health: A Study of Organic Agriculture* (1947 Devin-Adair; reprinted 2006, University of Kentucky), Sir Albert Howard describes composting as a means of replenishing the natural humus and restoring soil functionality.

2. The Paradox of Agriculture

1 Wendell Berry, "The need for a 50-year Farm Bill." *Farming Magazine,* 2012, 12(4):24.
2 Paul Harvey, "So God Made a Farmer," speech to the Future Farmers of America, 1978, replayed in a Dodge Ram truck commercial during Super Bowl XLVII in 2013.
3 US Environmental Protection Agency, 2009. Report EPA841-F-96-004A. Washington, DC.
4 Gibbs, S. et al. 2006. "Isolation of Antibiotic-Resistant Bacteria from the Air Plume Downwind of a Swine Confined or Concentrated Animal Feeding Operation." *Environmental Health Perspectives* 114(7): 1032–1037.
5 The video can be found at: cbsnews.com/video/watch/?id=3773487n&tag=related;photovideo (It's not pretty).

3. Farm Subsidies

1 Environmental Working Group, 2012. Farm Subsidy Database. farm.ewg.org
2 National Agricultural Statistics Service. Quick Crop Statistics. Table 10. "Harvested acreage and yield per acre (for Iowa, 2009)." USDA, Washington, DC.
3 *The Economist.* September 22, 2012. Agricultural subsidies. economist.com/node/21563323
4 Kaskey, J. August 13, 2009. Monsanto to Charge as Much as 42% More for New Seeds (Update3). Bloomberg.
5 Schueler, T. 1995. Site Planning for Urban Stream Protection. Center for Watershed Protection. Metropolitan Washington Council of Governments. Silver Spring, MD.
6 Fischer, J. and 11 others. 2008. "Should agricultural policies encourage land sparing or wildlife friendly farming?" Frontiers in Ecology and Environment. 6.
7 Mäder, P., A. Fliessbach, D. Dubois, L. Gunst, P. Fried and U. Niggli. 2002. "Soil fertility and biodiversity in organic farming." *Science* 296: 1694–1697.
8 MacArthur, R.E. and E.O. Wilson. 1967. *The Theory of Island Biogeography.* Harvard University Press, Cambridge.

9 See, Tilman, D., C.L. Lehman and C.E. Bristow. 1998. "Diversity-stability relationships: Statistical inevitability or ecological consequence?" Am. Nat. 151:277–282. Also, Pimm, S.L. and C. Jenkins. Sustaining the variety of life. *Scientific American* 293: 66–73.

4. Toward a Sustainable Agriculture

1 Brundtland, G.H. et al. 1987. "Report of the World Commission on Environment and Development." General Assembly Resolution 42/187, 11 December 1987, United Nations.
2 Based on a segment of the film, *Dirt: The Movie.* 2008. Common Ground Media.

5. Sustainable Meat — A Contradiction in Terms?

1 The Smallwood Group. 2012. Analysis of FAO (United Nations) findings. smallwood.com.au/charts.htm
2 Haney, S. April 30, 2012. "How much meat do we eat?" *The Economist, online.* realagriculture.com/2012/05. Haney uses FAO data to chart the annual intake of meat by the 18 largest consumer nations. Recently, the US Centers for Disease Control and Prevention presented an estimate about half the size of the FAO estimate, and the National Cancer Institute (of NIH) has estimated that consumption is about 25 percent of that presented by FAO. The latter, however, appears not fully validated at this time.
3 Willett, W.C. 1994. "Diet and health: What should we eat?" *Science* 264: 532–537.
4 Hunninghake H.B., K.C. Maki, P. O. Kwiterovich Jr, M.H. Davidson, M.R. Dicklin, and S.D. Kafonek. 2000. "Incorporation of lean red meat into a National Cholesterol Education Program Step I Diet: A Long-term, randomized clinical trial in free-living persons with hypercholesterolemia." *Journal of the American College of Nutrition* 19:351–360.
5 Duckett, S. K., D. G. Wagner, et al. 1993. "Effects of time on feed on beef nutrient composition." *Journal of Animal Science* 71: 2079–88.
6 As I write this, the U.S. Congress is debating a bill that would make documenting the mistreatment of livestock on farms, in feedlots, and at packing plants illegal. This will obviate the efforts of activists to prevent animal cruelty.

7 Chickens, being omnivores, require some grains in their diets, unlike cattle, which evolved to eat grass.

8 Savory, A. 1999. *Holistic Management.* Island Press, Washington, DC.

9 To be sure, there is an energetic debate underway about the superiority of rotational versus continuous grazing. The superiority of one strategy over the other may depend on the system, e.g., pasture vs. rangeland, being grazed. See, Briske, D.D., et al. 2008. "Rotational grazing on rangelands:Reconciliiation of perception and experimental evidence." *Rangeland Ecology & Management.* 61: 3–17, Also see, Teague, W.R., et al. 2011. "Grazing management impacts on vegetation, soil biota and soil chemical, physical and hydrological properties in tall grass prairie." *Agriculture, Ecosystems and Environment* 141: 310–322.

10 Carter, C.A. and H.I. Miller. 2012. "Corn for food, not fuel." *The New York Times,* July 30, 2012.

11 Data: NASS. 2007. Census of Agriculture. USDA; and USEPA. 2012. Ag 101: Major crops grown in the United States. epa.gov/oecaagct/ag101/cropmajor.html It should be understood that much of the nation's pasture land cannot converted to crop production; the quality of the soil will not support it.

12 Estimates range from 6.0 to 50.0 gallons per day (gpd), depending upon breed and lactational status.

13 The size of the US herd is 96 million cattle, 60 million beef cattle, according to NASS. (2007 Census of Agriculture, USDA); R. Plain. 2012. "Semi-annual cattle inventory." University of Missouri. agebb.missouri.edu/mkt/bull12c.htm

14 Reece W.O. 2004. "Kidney Function in Mammals." *In,* Reece W.O. (Editor), *Dukes' Physiology of Domestic Animals,* 12th (ed.), Cornell University.

15 Frank, D.A. 2008. "Ungulate and topographic control of nitrogen: phosphorus stoichiometry in a temperate grassland; soils, plants and mineralization rates." *Oikos* 117:591–601.

16 US EPA. 2013. Inventory of US Greenhouse Gas Emissions and Sinks: 1990–2011. EPA 430-R-13-001. Washington, DC.

17 Wightman, J. 2009. "Production and mitigation of greenhouse gases in agriculture." *Climate Change and Agriculture: Promoting Practical and Profitable Responses.* Cornell University.

18 DeRamus, H. A., T. Clement, D. Giampola, and P. Dickison. 2003. "Beef Production Efficiency in Forages and Grazing Management Systems as Monitored by Methane Emissions." *Journal of Environmental Quality.* 32:269–277.

19 Finishing is the final stage of lamb, beef, and pork production, when a layer of fat (which will tenderize and flavor the meat) is produced around the muscle.

20 See, Kleppel, G.S., C. Girard, E. LaBarge and S. Caggiano. 2011. "Invasive plant control by livestock: from targeted eradication to ecosystem restoration." *Ecological Restoration* 29: 209–211; Kleppel, G.S. and E. LaBarge. 2011. "Using sheep to control of purple loosestrife (*Lythrum salicaria*)." *Invasive Plant Science and Management* 4: 50–57; Kleppel, G.S., C.B. Girard and E. LaBarge. 2011. "Draft -Intensive Rotational Targeted Grazing: A Protocol for Ecosystem-Based, Invasive Plant Management and Habitat Restoration in New York State." Submission to New York State Department of Environmental Conservation; Caggiano, S. A. O'Connor and G.S. Kleppel. 2010. "The use of goats to control multiflora rose (*Rosa multiflora*) in a pasture in the Hudson Valley of New York State." Technical Report. 101230/02. University at Albany.

21 As I write this, Congress is considering making it illegal to photograph activities in an agricultural facility, further limiting the public's access to validated information about how their food is produced.

6. Diversity in Agriculture

1 Gabriel, D. et al. 2009. "The spatial aggregation of organic farming in England and its underlying environmental correlates." *Journal of Applied Ecology* 46: 323–333.

2 Ananda, R., May 17 2010. "Mark of the Beast: Obama's Latest Monsanto Pick, Elena Kagan." dissidentvoice.org/2010/05/mark-of-the-beast-obama's-latest-monsanto-pick-elena-kagan/

3 Powles, S.B. 2008. "Evolved glyphosate-resistance around the world: Lessons to be learnt." *Pest Management Science* 64:360–365; Heap, I. "The International Survey of Herbicide Resistant Weeds." Available at weedscience.org

4 Schafer, M.G., A.A. Ross, J.P. Londo, C.A. Burdick, E.H. Lee, et

al. (2011) "The Establishment of Genetically Engineered Canola Populations in the US." *PLOS ONE* 6(10).

5 Quist, D. and I. Chapella. 2001. "Transgenic DNA introgressed into traditional maize landraces in Oaxaca, Mexico." *Nature* 414: 541–543.

6 Billeter, R. et al. 2008. "Indicators for biodiversity in agricultural landscapes: a pan-European study." *Journal of Applied Ecology* 45:141–150.

7. Energy and the Future of Farming

1 Heinberg, R. 2003. *The Party's Over: Oil, War and the Fate of Industrial Societies.* New Society Publishers, Gabriola Island, BC Canada.

2 The simple sugars produced by photosynthesis are used to make nearly all of the other compounds (e.g., fats, proteins) needed by the plant.

3 Land Art Generator Initiative. 2009. Total surface area required to fuel the world with solar. landartgenerator.org/blagi/archives/127
Ikerd, J. 2008. *Crisis and Opportunity.* University of Nebraska Press, Lincoln.

5 A staff article in *Farm Futures* (June 7, 2013) states, "As many Americans adjust budgets to keep food on the table, a recent World Bank study says oil prices are the biggest driver behind food prices — and represent a reason to keep an eye on government policy." The study, by J. Baffes and A. Dennis notes that oil-driven price increases are permanent and susceptible to "spikes". Over the seven years studied, wheat stocks declined by 17% while oil prices rose by 220%. The study is available as a pdf download at bit.ly/1kzPpGf

8. The New Normal

1 From the film, *The Greenhorns.* Produced by Severine von Tscharner Fleming (2011).

2 The Barbers were knocked down but not out. The Barber Family picked up the pieces and began rebuilding. Today, they are once again up and running, stronger than ever.

3 The Disaster Center. disatercenter.com/newyork/tornado.html

4 Most climate scientists agree that the extent of climate change would be reduced if atmospheric CO_2 levels were maintained below 350 ppm (see, Hansen, J. et al. 2013. "Required reduction of carbon emissions to protect young people, future generations and nature." *PLOS ONE,* 8.)

5 Turner B.L. II and D. Lawrence. 2012. "Land architecture in the Maya lowlands: Implications for sustainability." In P. Gepts et al. (Editors). *Biodiversity in Agriculture.* Cambridge University Press, New York.
6 Diamond, J. 2005. *Collapse.* Penguin, New York.

9. The Emergent Market

1 Actually, I was wrong. The community came together and saved his farm.
2 Schlosser, E. 2002. *Fast Food Nation: The Dark Side of the All-American Meal.* Houghton-Miflin, New York, pp. 145–146.
3 Most agricultural products are sold in units of 100 pounds or hundred-weight.
4 US Department of Labor. Bureau of Labor Statistics. data.bls.gov/cgi-bin/surveymost?
5 National Family Farm Coalition. *Food, Inc.* and *Fresh: Facts and Solutions Needed to Fix the Food System.* Available at nffc.net/Learn/Fact%20Sheets/food%20inc%20and%20fresh.pdf
6 Top Class Actions reported: In January 2013, DFA agreed to pay $158.6 million to settle a separate price-fixing class action lawsuit accusing DFA, Dean Foods and others of conspiring to fix raw milk prices across the southeastern United States, short-changing dairy farmers. The settlement was in addition to a $145 million settlement reached in the same case (In re Southeast Milk Antitrust Litigation) in 2012. Available at: topclassactions.com/lawsuit-settlements/lawsuit-news/3789-dairy-farmers-of-america-to-pay-46m-price-fixing-settlement

10. The Consumer in a Changing Food System

1 Food Safety & Inspection Service US Department of Agriculture, April 12, 2011. fsis.usda.gov/factsheets/Meat_&_Poultry_Labeling_Terms/index.asp
2 To ensure that fully grown broiler chickens are not crowded we use a simple formula: Number of chickens divided by floor area in the coop must be at least two square feet.
3 Although the price of meat from the industrial system remains lower than meat produced from small-scale grassfed operations, the price has

risen significantly over the past decade. The rise in the price of industrial meat is most likely due in large part to the rising cost of fossil fuels. In January 2004, the "lowest" price for beef nationally was $1.07/lb. In December 2013, the price was $1.86/lb, a decadal increase of 73.7% (UDSA Market News, Feb 14, 2014 and Index Mundi and indexmundi.com/commodities/?commodity=beef&months=120)

4 World Health Organization. 2013. "General information related to microbiological risks in food." who.int/foodsafety/micro/general/en/index.html

5 USDA. Food production daily. 2008; US CDC 2007

6 Reynolds, G. 2007. "FDA food safety problems blamed on lack of funding." *Food Production Daily.* 25 April, 2007

7 Dairy Lobbying. OpenSecrets.org reports that in 2012 the dairy industry spent $7,222,249 to lobby Congress. The largest contributors were the International Dairy Foods Association ($1,616,000), Land O' Lakes (1,200,000), Dairy Farmers of America ($1,122,000) and Dean Foods ($730,000). Data were available for downloading on July 29, 2013 at opensecrets.org/lobby/indusclient.php?id=a04&year=2010

8 Morgan D., S. Cohen and D.M. Gaul. "Dairy Industry Crushed Innovator Who Bested Price-Control System." *Washington Post,* December 10, 2006.

9 Reading Terminal Market, Philadelphia, PA, readingterminalmarket.org/

10 Pike Place Market, Seattle, WA, pikeplacemarket.org/

11 GrowNYC — NY Green Markets System, New York, NY, grownyc.org/

11. Slow Money

1 Trasch, W. 2010. *Slow Money.* Chelsea Green, White River Junction, VT.

2 For a description of worker abuse in the meat packing and fast food industries see Schlosser, E. 2002. *Fast Food Nation.* Penguin.

13. Scaling Up — How Local Farms Will Feed America

1 Peters C.J., N.L. Bills, A.J. Lembo, J.L. Wilkins JL and G.W. Fick. 2009. "Mapping potential foodsheds in New York State: A spatial model

for evaluating the capacity to localize food production." *Renewable Agriculture and Food Systems* 24:72–84.

2 Pirog, R and A. Benjamin. 2003. "Checking the food odometer: Comparing food miles for local versus conventional produce sales to Iowa institutions." Leopold Center for Sustainable Agriculture, Iowa State University, Ames. Available at: leopold.iastate.edu/sites/default/files/pubs-and-papers/2003-07-checking-food-odometer-comparing-food-miles-local-versus-conventional-produce-sales-iowa-institution.pdf

3 Smith, A. and J.B. MacKinnon. 2007. *Plenty: Eating Locally on the 100-mile Diet.* Three Rivers, New York.

4 Asami, D.K. et al. 2003. "Comparison of total phenolic and ascorbic acid content of freeze-dried and air-dried marionberry, strawberry, and corn using conventional, organic and sustainable agricultural practices." *Journal of Agricultural and Food Chemistry* 51: 1237–1241.

5 Vajello, F. et al. 2003. "Health-promoting compounds in broccoli as influenced by refrigerated transport and retail sale period." *Journal of Agricultural and Food Chemistry* 51:2029–3034.

6 Schwartz, J.D. "Buying Local: How it Boosts the Economy." *Time Magazine.* June 11, 2009. content.time.com/time/business/article/0,8599,1903632,00.html
"Buying local means more money stays in your community." Examiner.com March 11, 2010. examiner.com/article/buying-local-means-more-money-stays-in-your-community
Sacks, J. 2002. "The Money Trail: Measuring Your Impact on the Local Economy Using LM3." New Economics Foundation, London.

7 National Farm-to-School Network. Available at: farmtoschool.org/

8 Breslin, K. 2012. "Farms on Prisons will Reduce National Prison Budget." Policymic. policymic.com/articles/8869/farms-on-prisons-will-reduce-national-prison-budget

9 James Jiller. "Doing Time in the Garden — Life Lessons through Prison Horticulture." *New Village Press*, 2006.

10 Boeri, D. 2011. "Life On 'The Farm:' Lawrence Jail Prepares Inmates For Re-Entry." WBUR, National Public Radio, Boston, MA. wbur.org/2011/04/12/prison-re-entry

11 The Urban Farming Global Food Chain. urbanfarming.org

12 Allen, W. 2012. The Good Food Revolution. Gotham, New York.
13 Sky Greens. 2011. skygreens.appsfly.com/products

14. The Emergent Agriculture

1 The concept of land health is expressed poignantly in a series of essays by Aldo Leopold, published posthumously in a book entitled *For the Health of the Land,* edited by J.B. Callicott and E.T. Freyfogle (1999. Island Press, Washington, DC).

Index

About the Author

GARY KLEPPEL IS PROFESSOR OF BIOLOGY and, director of the graduate program in Biodiversity Conservation and Policy at the State University of New York at Albany. He received a PhD in Biology from Fordham University in 1979, and studied marine ecosystems early in his career. His current research deals with sustainable land use, agriculture and the ecology of human-dominated landscapes.

Kleppel and his wife operate a small farm about 15 miles west of Albany where they produce grass-fed lamb, wool, free-range poultry, cage free eggs, and artisan breads that are marketed directly to the public. The Kleppels use methods of food and fiber production that sustain the ecological integrity of the land. Energy is provided by photovoltaics, the only inputs used on the farm are composts, and pastures are managed for soil health and biodiversity.

New Society Publishers

ENVIRONMENTAL BENEFITS STATEMENT

New Society Publishers has chosen to produce this book on recycled paper made with **100% post consumer waste,** processed chlorine free, and old growth free.

For every 5,000 books printed, New Society saves the following resources:[1]

20	Trees
1,772	Pounds of Solid Waste
1,949	Gallons of Water
2,543	Kilowatt Hours of Electricity
3,221	Pounds of Greenhouse Gases
14	Pounds of HAPs, VOCs, and AOX Combined
5	Cubic Yards of Landfill Space

[1]Environmental benefits are calculated based on research done by the Environmental Defense Fund and other members of the Paper Task Force who study the environmental impacts of the paper industry.